D1256058

American Windmills

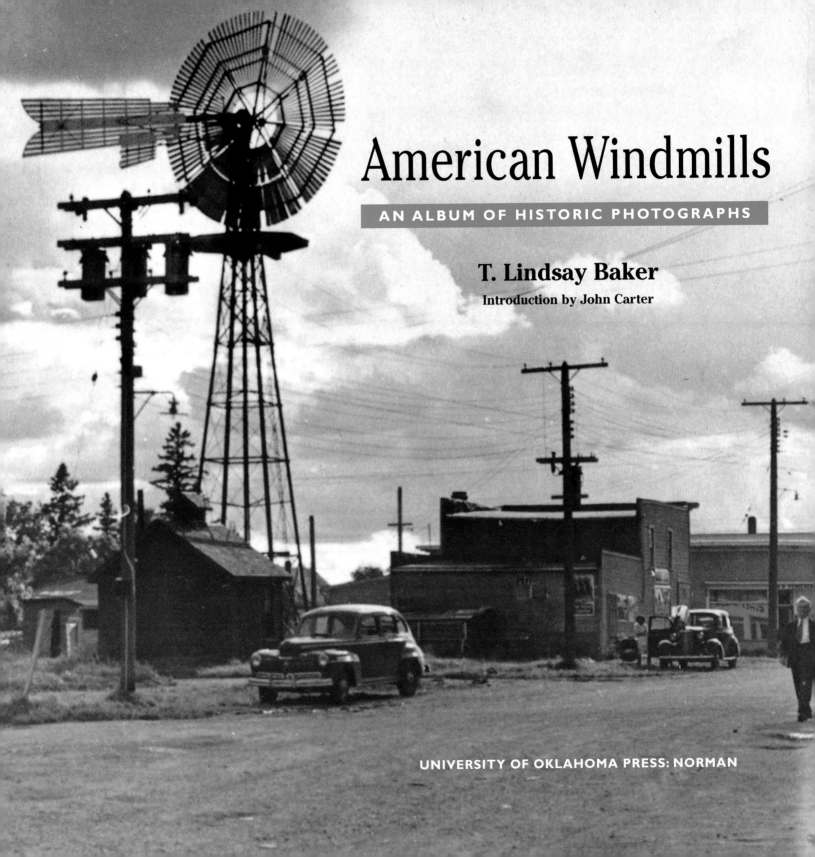

# American Windmills

## AN ALBUM OF HISTORIC PHOTOGRAPHS

**T. Lindsay Baker**

Introduction by John Carter

UNIVERSITY OF OKLAHOMA PRESS: NORMAN

This book is published with the generous assistance of the Olive B. Cole Foundation.

LIBRARY OF CONGRESS CATALOGING-IN-PUBLICATION DATA

Baker, T. Lindsay.
    American windmills : an album of historic photographs / T. Lindsay Baker ;
    introduction by John Carter.
        p.  cm.
    Includes index.
    ISBN 0-8061-3802-5 (alk. paper)
    ISBN 978-0-8061-3802-2
    1. Windmills—United States—Pictorial works.  I. Title.
    TJ825.B258  2007
    621.4'530973022—dc22                                                    2006025550

The paper in this book meets the guidelines for permanence and durability of the
Committee on Production Guidelines for Book Longevity of the Council on Library
Resources.

1  2  3  4  5  6  7  8  9  10

*In memory of B. H. "Tex" Burdick, Sr.*

# CONTENTS

# Preface

Idon't know when I purchased my first old photograph of a windmill in a junk shop; I probably paid no more than a dollar for it. I do know, however, when I started copying them. It was in 1974 that I took a black-and-white photography class from Herschel Womack in the journalism department at Texas Tech and bought my first single-lens reflex camera. Herschel taught me how to use screw-on close-up lens adapters and I started copying old windmill photographs wherever I could find them. From then on, my search for historic windmill pictures became unstoppable. In time I gathered hundreds of them, each carefully sleeved and identified.

I discovered that scores of public repositories and private collections contained historic photographs of windmills. Where possible, I secured reproductions of the best of these images, seeking visual documentation of these machines that were familiar and almost ever-present throughout rural America. My great delight came in finding a first photograph of a particular brand or production model. I also searched for examples that captured on film the multiple ways people attempted to use free wind power.

This volume represents a certain plateau in my search. Editor Charles E. Rankin and I looked at approximately 2,000 photographs to select the ones that appear on these pages. The decisions were never easy. For each image we chose, there were multiple competitors.

The work of many people contributed substantially to this effort. More than anyone else, I recognize the late B. H. "Tex" Burdick, Sr., of El Paso, Texas. From the 1920s through the 1940s, he photographed his employees doing all the many aspects of windmill erection and maintenance, creating a visual record unequalled by any other known American collection. In 1979–1981 Burdick allowed me to copy his collection onto safety film and place them in the Panhandle-Plains Historical Museum, where this remarkable record is preserved today.

Another important source was corporate archives of firms that made and used windmills. The Baker Manufacturing Company, the Union Pacific Railroad, and Southern Cross Pumps and Irrigation, Pty. Ltd. permitted use of their images in this book.

John Carter, senior research associate at the Nebraska State Historical Society, has given me encouragement for a quarter century. In the summer of 1977, we met in Lincoln and spent hours poring over the magnificent images made in the 1880s through the 1900s by Solomon D. Butcher, preeminent recorder of the sod house frontier. John helped me understand that "windmill pictures" contain more important subject matter than the wind machines themselves, and his insights greatly shaped my subsequent studies. He very generously agreed to write the introduction to this volume.

Many other archivists and librarians have gone to great lengths to search out photographs from their repositories. Claire R. Kuehn, Lynnette Guy, Lisa Lambert, and Betty Bustos, all at the Panhandle-Plains Historical Museum, top the list from the institution that holds more of these images than any other. At the Texas Tech University Southwest Collection, Janet Neugebauer, Roy Sylvan Dunn, and David Murrah greatly facilitated my efforts to find important windmill photographs. Jim Nottage, as curator of history at the Kansas State Historical Society, also pulled out all the stops to be helpful as I worked in files at this extraordinary repository. Curator Carla Hill presides over the finest collection of photographs that record historic windmill manufacture at the Batavia Historical Society in Illinois, and she did everything possible to assist me as an out-of-state researcher. As the pictures credits show, other collections assisted substantially, among them the San Diego Historical Society, the U.S. Geological Survey Photo Archives, the Library of Congress, the University of California at Davis, the California State Library, the Plymouth (Michigan) Historical Society, the

Colorado Historical Society, the Kregel Windmill Company Museum, Fort Concho National Historic Landmark, the Chicago Historical Society, the Canadian National Historic Windmill Centre, and the Archives of Manitoba.

Private individuals were generous as well with copies of historic photographs in their possession. These included A. Clyde Eide; J. Kenneth Major; Robert M. Voegel; George A. Carlson; Frankie Jo Rintoul; B. H. "Tex" Burdick, Sr.; and David Huber.

Private dealers have provided me with opportunities to purchase some magnificent photographs. Among these have been B. William Henry, Guy Peaudecerf, Craig Donges, Helen and Garnet Brooks, Virgil L. Nelson, Ed Burgi, and David Schnakenberg.

My wife, Julie, spent hours editing the manuscript and "reading" the photographs for content, greatly enhancing the quality of the completed work.

Finally, I want to express my gratitude to the long-gone photographers, known and unknown, who took many of the pictures in this book. It is through their efforts that we enjoy the visual record of windmills and their use in the decades past.

T. Lindsay Baker
Rio Vista, Texas

American Windmills

A crew from the Burdick & Burdick Company assembled this thirty-foot steel tower for a sixteen-foot Challenge 27 windmill made by the Challenge Company. The site was on Dr. Francis Cole's ranch property west of Las Cruces, New Mexico. As his men used cables to raise the completed tower up toward its vertical position over the well, B. H. Burdick, Sr., stopped the action with his Kodak No. 2 folding camera. Circa 1931–1932. Courtesy of B. H. Burdick, Sr.

# Introduction

*John Carter*

Every profession has its little tricks, the ones taught by experience. Those of us who study culture on America's Great Plains by getting out and looking at it have a doozy: Just look for something vertical.

This plains landscape is very, very big and relentlessly horizontal. For those unaccustomed to it, the experience can be numbing, deluding us into the belief that there is no grand geography to be seen here! Truth is, this landscape is so grand that it simply can't be seen in one easy eyeful. It is not like mountains, which seem majestic because you can see them in one convenient eyeful. The intrinsic grandeur of the Great Plains is concealed by the camouflage of the very large. On this land, human beings really are but fleas on the back of an elephant.

So out on that grand and vast landscape, where one field of vision is forced to overlap another, if something sticks up you can be pretty sure that it is important. That vertical thing marks a spot where something got different, a point of transition. That transition creates significance. So in this geographic drama, verticality is a leitmotif. Let me offer some examples.

Those traveling the Overland Trail regularly noted a landmark near present-day Central City, Nebraska: "The Tree." It was a solitary cottonwood, visible for twenty miles, passed by countless Mormons, '49ers, and other westering souls. So notable was this tree that the initials carved into the giant trunk by scores of passersby quite literally whittled it to death. Lone Tree actually became the name of the place. Think about the absurdity of that in almost any other geography short of a desert. And 350 miles later those same pilgrims nearly wept as they came upon Chimney Rock, the first geological feature to poke its rocky nose skyward.

The Euro-Americans who came to the plains and stayed found the absence of trees unbearable, and planting trees was among the first things they did. This was patently absurd, as here young trees require lots of work. But plant they did.

And just how much work was it to get trees to grow? Think about this: There isn't much in the way of rainfall, particularly west of the 100th meridian, and it takes trees upward of three years for their roots to reach groundwater. The hot winds of summer rapidly turn unwatered saplings into matchsticks, so once or twice a week you need to pump water and dump it on those little sticks in the ground.

Experience quickly teaches that when you regularly dump water on prairie ground, appreciative grasses grow up, competing with the tree for sunlight. That requires the tree planter to take hoe in hand and chop down the rival flora.

All of that work, three years of watering and hoeing, for something that won't even produce kindling for a decade. But it is worth it, for those trees, those vertical trees, bring that vast endless horizon into an understandable human scale.

Often a solitary grove on the prairie terrain is the site of a failed dream, a homestead gone bust. The buildings melted away long ago, but the carefully planted trees remain. And in rural cemeteries, ranks of coniferous trees guard and define the formal rows of headstones, a quiet memorial to the loved ones who lie there.

When trees do mature they are transformed into another vertical. A simple row of posts connected by barbed wire artificially draws a line defining the geography of a ranch or farm. No such delineation of the land existed before human imagination and hands enforced one upon it. From a road, a break in a fence is a ritual entrance to the intimate spaces of a rural geography where houses and outbuildings precisely define and direct our understanding of every particular space.

The spire of a modest rural church, which can be seen for miles in all directions, is monumental, far more so than a similar structure lost in the muddle of close-pressed urban architecture. Prior to being hidden by the now-mature urban forests, the buildings of cities and towns on the plains literally exploded up off the landscape.

The point is that the vertical creates an ontological distinction on the plains, creating a place where the work of human hands inserts a new meaning. It is, therefore, a good place to find human drama and poetry. The phenomenon of the windmill is a case in point.

Windmills are, like trees, monuments to a hoped-for permanence, and as icons they quickly transcend their mechanical purpose of delivering water. For nearly three decades I have studied thousands of photographs in which the windmill is the central feature, and that centrality is no accident. Because they literally tower, they have the advantage of scale and form over other vertical elements, such as skinny utility poles and fences.

Look at the form. When I read diaries and reminiscences, I find recurring but strained comparisons of windmills to flowers. In the sense of a

*Winter weather presented maintenance problems for the users of windmills, for it froze up improperly insulated water systems. Here a Dempster No. 4 Vaneless windmill made by the Dempster Mill Manufacturing Company of Beatrice, Nebraska, was "snowed in" on a farm near Minot, North Dakota, about 1925. The handwritten caption on the back of the picture reads "Walter's House up west." From the author's collection.*

*From the second floor of a house, someone hastily snapped this picture of a tornado approaching windmill-filled Mullinville, Kansas, on June 11, 1915. From the author's collection.*

Rorschach test one might see a giant sunflower, but then one can also see Elvis's head in a potato. The geometry of a windmill is hard and unyielding; that is the idea. A windmill waving gently in the wind just wouldn't work very well, would it? No, much as we would like to see windmills as a part of the organic landscape, they are anything but. They were birthed in the dark and dirty and disorderly world of nineteenth-century factories. They were ground and pounded from metal and hewn from cut lumber.

Even in the elegance of the wooden-wheeled basket windmill that opens and closes, blossom-like, out of respect for wind velocity, there is nothing natural. It is strictly industrial mechanics. What grace and beauty windmills possess come not from the workings of nature but rather from the craft and skill of human engineering.

Symbolically, the windmill calls to mind monuments: the obelisks, like those found in graveyards or the Washington Monument, for that matter, or the Trylon and Perisphere of the 1939 New York World's Fair. With the former, a symbol drawn from ancient Egypt, comes a sense of permanence and with the later, an optimism for the future.

There is an esoteric understanding of these devices as well. Their function is alchemy—to take a powerful natural energy source, the wind, and turn it into water.

For many urban dwellers, the windmill loses that association with wind and water. The device becomes a near-icon, an image appearing on a package of bacon or a carton of butter. In the aisles and checkout lanes of supermarkets, distant as they are from the actual origins of the products they sell, the shape and form of the windmill is a mere curiosity, not unlike agriculture itself.

During World War II, Nebraska was home to eleven army airbases that were used as training facilities and pit stops for planes being ferried to one coast or the other. In November 1944, two servicemen were taking a P-38 from Kansas City to Denver and met disaster.

As they flew at night, their plane developed serious engine trouble, forcing them to bail out. The plane crashed, but the airmen parachuted safely to earth. They alighted on the Seth Hannah ranch in Cherry County, Nebraska, an immense and sparsely populated sand hills county that is larger than the state of Connecticut. They soon found that their troubles had just begun.

One of the two men suffered a seriously sprained ankle and they were lost in the dark, a long way from any town. Ranchers in the area later reported hearing an explosion, but because the plane did not burn, there was no way to locate the downed flyers at night.

The injured flyer was only able to crawl on his hands and knees, which, in a land filled with cactus and sandburs, would be like crawling through broken glass, thumbtacks, and razor blades. So the two separated, and the uninjured man went in search of help. At first light the next morning Mr. Hannah found the uninjured man, who had spent a long night walking in circles.

Exhausted and dehydrated, the pilot's first words to Hannah were, "Do you have a drink of water with you?" The rancher pointed to a windmill and well about fifty feet away. "My God," exclaimed the flyer, "I ran into six or seven of those things during the night and wondered what they were." The newspaper article that described the rescue simply explained that the wandering airman was from Alabama.[1]

One has to feel for the terrified young man who was lost in a very alien environment heroically trying to find help for his comrade, who was indeed rescued later. Yet the irony of a man being so close to water and not recognizing the device that produces it is jarring, nearly to the point of comedy.

So our effort to ferret out where the windmill actually fits in a larger iconography is handicapped at the beginning by the fact that a large portion of our population is concentrated in urban America and distant from the landscape for which the windmill serves as an icon. Separate the windmill from its landscape and you separate it from its meaning. My friend the poet and novelist Jim Harrison observed on numerous occasions that of all of the crimes committed by Hollywood against the American people, the most egregious is convincing us that all stories are geographically interchangeable. He is, of course, right; plains stories are quite different from those of the mountains, forests, rivers, or oceans. For the most part, the windmill is a Great Plains story.

In the same sense that there is an esoteric meaning to windmills as they relate to place, there is likewise an esoteric meaning as they relate to people. Being a person terrified of heights, I have only twice in my life completed the white-knuckled task of climbing a windmill tower. That weakness does, however, provide me with a particular admiration for those who erect and maintain these things. Not only do they climb those towers, but they use their hands to *work* up there. And they do so without the complicated safety riggings and harnesses that I would demand.

So when I look at the faces in the photographs of the daredevils working on windmills, I understand the pride in profession reserved for those who dare to go where angels fear to tread. Angels, after all, have wings.

In that regard, there is one photograph, a beautiful photograph, that is just plain awe-inspiring! It is a bird's-eye view of San Diego (and for the record, San Diego is not situated on the Great Plains) with a windmill in the immediate foreground. Perched atop its tower are two finely dressed young ladies.

The photograph is heroic in both scale and subject. I find it unsettling and amazing because somewhere, just a little above the windmill in the picture, stands a slack-witted photographer who is on a *taller* structure with a wind-catching view camera the size of a microwave oven, head shrouded under a dark cloth, waving around a large glass plate-bearing negative carrier, another wind-catching device.

More than that, because view cameras lack the image-correcting prisms of modern single-lens reflex cameras, the photographer is looking at this dizzying image upside down. It does make you wonder just what this person wouldn't do for a picture, and that leads us to an important question: Why do photographers take pictures of and about windmills?

As often as not, the answer to that question is simple: They were paid to take them. But as with most things in life, there is more to it than that.

I have argued often that for a number of reasons, photography is distinct in the visual arts in that it demands the photographer's physical presence. A skilled painter can produce a painting from memory or imagination, any time, any place. The photographer has to be there, and be there when the event being photographed occurs.

That person also has to select precisely what to photograph. There is the old saw that photographs don't lie, and before the digital age, that was pretty much true. Whatever is set before the lens appears in the picture. This again distinguishes photography from other forms of two-dimensional representations of reality. A person who draws a scene can choose its contents; the person who photographs it must take what is there. So the photographer faces a critical decision: What, of the infinite universe of potential views available at a given time and place, do I actually capture on film?

Several things shape what the photographer chooses at that moment. The obvious starting place is that person's visceral response to place and event. But that is just a starting place. Like any visual art, photography anticipates an audience, a very specific audience. There may actually be some people who take a picture solely for the joy of taking it and then never show it to a living soul. But for the most part, the ultimate intention of the photographer is to have someone else see the picture and understand it.

*These two couples rode around the Erick, Oklahoma, countryside in horse-drawn top buggies at the turn of the twentieth century. They took a break at the base of this regular-pattern Eclipse windmill made by Fairbanks, Morse and Company so their animals could take a drink. From the author's collection.*

Thus, the expectation of that audience has quite a lot to do with the picture that is taken. Proud parents photographing a two-year-old understand their intended audience (grandparents, aunts and uncles, the occasional hapless friend), as does the photojournalist at a crime scene or, for that matter, the pornographer. The photographer intends and the audience expects. These are the first two elements of the photographic transaction.

Human subjects have something to say about how the picture looks, too, and that is the third element of the transaction. Witness the two ladies mentioned earlier. Imagine how different the picture would read if they were clad, say, in riding gear or dressed as servants in a wealthy household or clothed in Salvation Army uniforms. Everything else in the picture would stay the same, but the meaning would shift dramatically.

The subject, like the photographer, senses the expectation of the audience. A championship bowling team poses one way, a family another. How different a high school graduation portrait looks from that teenager in action; how different the fiftieth-wedding-anniversary portrait from the couple in their everyday activities.

So to understand a given photograph or, in the case of this book, a universe of images assembled on a theme, one has to understand the process of the interaction of time and place, the photographer's intention, the subject's intention, and the audience's expectation. It is an interaction of people at a specific place at a specific time. Their effect and their ultimate meaning is derived from this interaction.

Another key to understanding the deeper meanings of photographs is philosophical. Ontologically, photographs are simply an amalgam of animal, vegetable, and mineral. Paper, made most often of cotton or wood fiber, is the vegetable part. Light-sensitive silver, the material that creates the actual image, is the mineral. And the emulsion that carries the silver—the gelatin, made of boiled hides, bones, and tendons—is the animal.

These three elements come together with startling effect. With the intervention of a lens they reproduce reality perfectly. The real nature of a photograph is no different than any other kind of picture; it is simply a two-dimensional abstraction. But photography is such a convincing abstraction that when it is viewed it is taken for the reality itself. When someone asks "Do you want to see my kid?" they are not thinking "Do you want to see a two-dimensional abstraction of my kid?" Rather, they are thinking that you are going to see the real deal.

This visual sleight of hand is important because it allows photography to not just represent reality but to cheat it. The nature of the real world is that time moves and space is fixed. Photography creates a frozen cross-section of time and allows space to move. This book is about that ontological cheating. It compares windmills through time and space, bringing views from the 1870s and the 1970s together and comparing those in North Dakota with those in Texas.

When we, as the viewers of these photographs, want to understand them properly, we must ask simple but important questions. First, what was the intention of the photographer? Why, out of an unlimited possible universe of views, did the photographer choose this view at this time? What are the icons, the visual symbols that the photographer chose to communicate with us? If there are human subjects in the photograph, what was their intention? Finally, what is the expectation of their intended audience?

Let me begin with the work of Solomon D. Butcher, a photographer heavily represented in this book and one with whom I am quite familiar. The work of Solomon D. Butcher, as Dr. Baker notes, is an important resource for understanding the settlement era in nineteenth-century America. Beginning in 1886 and inspired by a vision of a pioneer history recorded while that very history was being born, Butcher set out with both camera and notepad. He would pursue his project, an obsession really, in fits and starts for nearly twenty-five years.

With Butcher we know the intention of photographer and subject and the expectation of the audience. Butcher's pitch was quite different from that of other photographers of his time. He was not selling a photograph for you to send to the relatives back home. Rather, he was soliciting you to be a part of his grand history of Custer County, Nebraska. Both photographer and subject knew that decades later, we would be looking back at them. Thus, each individual image has an implicit narrative: of pride in having land of your own and hope for that land to flow with milk and honey.

In 1912, Butcher sold his collection to the Nebraska State Historical Society, which is also significant. He understood that there was a collective narrative in his work, and that body of work immediately became the visual metaphor for the settlement epoch in America. There is none other like it.

In these photographs there are nearly eight hundred that show windmills. Rather than tell you about the photographs, I invite you to look at them. The entire Solomon D. Butcher Collection held by the Nebraska State Historical Society is available on the Library of Congress's prestigious American Memory website.[2]

The sod house photographs for which Butcher is best known follow in a tradition known as the house portrait. House portraits are about estate, about having land, about having a family. The element that distinguishes the house portrait from a photograph of a house is that it is always populated. It is a photograph of people *and* their estate.

There is an unmistakable pride expressed in these photographs, a pride in accomplishment, a pride in being landed, and a pride in loved ones. I find these photographs remarkable in that because they are about hearth and home, by all rights, they should be cloying. But they are not. There is pride in good fortune but not hubris. I think we all understand that.

A close cousin of the house portrait is the farm or ranch portrait. These focus less on the home and more on the enterprise of farming and ranching. They emphasize outbuildings, livestock and equipment, and, yes, windmills. They are cousin to the occupational photographs that we will discuss later.

With both the house portrait and the farm or ranch portrait, the intended audience is limited and intimate: mostly family and maybe a few close friends (with, of course, Solomon Butcher's idiosyncratic adaptation of these genres for the purpose of historical narrative). They are rather like a letter, something recorded but intended for few eyes.

Where house portraits and farm or ranch portraits focus on the family, town portraits look at community. There is a similar element of pride, but that of a booster rather than that of a proud parent.

With town portraits, too, the audience shifts. These are not intimate glimpses but very public ones, often intended for an audience of strangers. This is not a difficult inference to draw; the fact that often the name of the town is written on the face of the image tells us something. Undoubtedly this was not done for the residents, as that would be like writing your own mother's name on your photograph of her.

The heyday of the town portrait unquestionably was the era of the postcard, from about 1905 to 1920. Postcards are that hybrid communication that combines imagery with the written word. While postcards are written and sent to individuals, their communication is de facto very public. You could hardly send a card expecting the letter carrier not to read it.

The town portrait can be a single view of the community or, as is often the case with postcards, a collection of images. This collection has a rather tight visual language that advances the image of the community as progressive, economically and socially stable, and moral. These pictures include schools, churches, banks, stores, railroad depots, and that town windmill.

What you don't see is likewise telling, no jails (jails suggest crime), no lawyers' offices (crime and lawsuits), no doctors' offices (sickness and accidents). They are not about the reality of the town but about the ideal image of the town.

The house portrait, farm or ranch portrait, and town portrait do share one common feature in terms of this book: they are photographs *of* windmills. The device is part of a visual ensemble that speaks of permanence, prosperity, and faith in the future. There are other photographs here, however, photographs that are *about* windmills.

Occupational photographs are akin to the farm or ranch portrait and certain elements of town portraits. They are both esoteric and exoteric: esoteric when they are intended for the work group, the employees, and exoteric when intended for audiences outside, for use in advertising or annual reports.

*Mike Sturm and his family gathered for a picture by Solomon D. Butcher in front of their square house and Dempster No. 1 Vaneless windmill in 1903 near Kearney, Nebraska. This model, from the Dempster Mill Manufacturing Company, Beatrice, Nebraska, was easy to identify from a distance due to the horse-shaped iron counterbalance weight. Courtesy of the Nebraska State Historical Society, Lincoln, Nebraska.*

While here it is not a straightforward process to infer the audience simply from the visual evidence, one can give it a shot and often come close to the mark. Some photographs are plainly intended for product promotion. You can spot these because the product names are writ large all over the image, often in an exaggerated form.

Workplace photographs, those shot in the offices and shops of a manufacturer, are harder to read. They are either formal, with everyone posed at their workstation but not working, or they are staged, with the staff pretending to work. One suspects that these are intended for newsletters, reports to shareholders, or for purposes of wholesale marketing. The message, of course, is that the staff is both skilled and hardworking and thus the product is of high quality and marketable.

Like most group portraits, the intended audience of an employee group portrait is the group itself. Whether formal or informal, these photographs are reminders of good times and prosperity. One would not expect such a photograph of, say, the employees who were recently laid off or the plant just before it shut down. We can be especially confident that the photographs

*A group of foundry workers, many of them wearing long protective aprons, stood for this picture outside one of the factory buildings of the Challenge Wind Mill and Feed Mill Company. These men made their living lifting heavy ladles of molten iron and pouring the red-hot metal into molds to make cast-iron parts for windmills and other products. The Fox River Valley of Illinois had many immigrants from Scandinavia, Eastern Europe, and Southern Europe. Courtesy of the Batavia Historical Society, Batavia, Illinois.*

that show staff members drinking at a company picnic or a convention are not intended for annual reports or for purposes of marketing.

Photographs taken for marketing purposes are both *of* and *about* windmills. By far the least interesting of this genre are the product shots. Their sole reason for existence is to show the product; no one looks at them for pleasure any more than one devotes recreational reading time to the operating instructions for kitchen appliances. They are unadorned and utilitarian.

But product displays at fairs are something else indeed! Fairs are the trade shows of the agricultural community and combine a sense of the carnival with a large scale that is truly awe-inspiring. I am especially taken with fairs in which huge numbers of windmills stand together like mechanical forests. You just have to marvel at the time and effort that went in to creating such a dazzling visual treat. Yes, they are just a product, but at the fair they become so much more.

Photographs become enduring visual benchmarks because of that trick of being able to cheat time. For that reason I like the photographs that show windmills being erected because if, as I would argue, these devices serve as both landmark and monument, then the moment when the landmark and monument is created is certainly an important transition, and that is when cameras come out.

The very act of constructing a windmill is monumental. Even with heavy equipment the task is considerable; without that equipment, the undertaking verges on the heroic. The construction process has two discreet elements: putting a tower up in the air and a pipe down in the ground. With my distaste for high places, I can assure you that I have no experience whatsoever with pushing the equipment skyward, but I can speak to the part about sinking a pipe to reach groundwater.

Back in 1976, I helped Roger Welsch, the author and former essayist for *CBS Sunday Morning*, install a pitcher pump on his farm on the Middle Loup River near Dannebrog, Nebraska. Because his farm is on the river, ground water was near the surface, at a depth of about ten feet. We were both young and naïve (some would say stupid; I won't quibble), and thus armed with ignorance we decided that we would simply drive a sand point down the required distance, creating what we came to learn is called a driven-point well.

For those of you unfamiliar with the fine points of the process, a sand point is a piece of two-inch pipe with a conical point at the end, like a rather fat javelin. The pipe has a length of screen that allows water in but filters sand and other debris out. The back end of the sand point is threaded to accept additional lengths of pipe to be added as the point is driven farther into the ground.

We decided to use a sledgehammer for the task of driving the point into the ground, and that taught me an important lesson: Never own a sledge-hammer, because you might be tempted to use it. We started off with an enthusiasm that soon plummeted into despair. In the early stages of our walloping, the point descended briskly. But soon the compounding realities of compression and friction became factors, and we pounded and pounded but seemed to be getting nowhere.

Roger decided to measure and see just how far we were advancing with each blow, so he placed a board alongside the pipe and drew a pencil line on it. I then gave the pipe a good smack and he repeated the process, drawing another line. The distance between the two lines measured one-eighth of an inch! It didn't take a mathematical genius to figure out that we were going to have to swing that sledgehammer nearly 1,000 times before we would get a single drop of water. Though we were thoroughly disheartened, we pressed ahead.

We did finally reach water, and when finally we pumped up the first gallon the elation was overpowering. It didn't take many tellings of that story before those wiser than us offered all of the creative ways that other generations had devised to solve the problem that we tackled with brute force, but to this day I hold a singular admiration for those who drove, bored, or dug wells.

Mealtime photographs have their own significance. We photograph meals that are more than just meals. If you scan through the shoebox of your family photographs, you will see what I mean. We photograph the two great national meals, Thanksgiving and Christmas. We photograph banquets. We photograph wedding feasts, birthday parties, and family reunions. That the photograph of a meal exists tells us that the meal was about more than just eating.

I am pretty sure that is the case with the photographs of the construction crews chowing down. If I am right and the act of erecting a windmill is significant, then the fact that photographs of the noon meal exist points dramatically to that significance.

In snapshot photography, people often pose in places of power. How many vacation photographs have we seen with folks standing erect in front of Mount Rushmore or with Old Faithful spewing up out of their heads? So, too, with pictures of people posed in front of windmills.

For some reason, people will do things in front of a camera that they would never ever do otherwise, and these photographs form their own genre. Of all of the truly astounding photographs in this book, the one with Corinne Baker regaled *as* a windmill takes the cake! I suppose if one were to be honest, it belongs with the other promotional photographs, probably best lodged with the photographs of windmills at fairs, but to me, this image transcends simple promotion.

I seriously doubt that anyone bought a Perkins Windmill from Mr. Titus because of this photograph. In fact, I think it likely that Mr. Titus gave this photograph to his customers after they had purchased their Perkins; his intended audience was a happy customer. If that be the case, is not Corinne Baker really a muse or perhaps even a goddess?

Silly as that might seem, I find it appropriate. If there is anything on earth that we could consider eternal, it is the wind. The wind is the embodiment of the spirit, the air we breathe, the soul. The water that the wind brings to us both sustains life and cleanses. Perhaps seeing windmills with metaphysical eyes is taking things a bit far. I don't really think we cast our gaze on a windmill or, in this case, the photograph of a windmill and find ourselves wrapped in spiritual reverie. On the other hand, when it is all said

*A gigantic Halladay and Wheeler's Patent Windmill made by the U.S. Wind Engine and Pump Company of Batavia, Illinois, ground grain at the Bennet Mill of Bennet, Nebraska, during the 1870s and 1880s. The miller wrote to the manufacturer, "I have worked at milling eighteen years, in as good water and steam mills as there are in the country, and I can make as good flour in this mill as in any of the same size and capacity I ever worked in." Courtesy of the Nebraska State Historical Society, Lincoln, Nebraska.*

and done, there is something about a windmill to which we resonate poetically. It *is* something symbolically greater than its physical form or mechanical purpose. If we didn't see and understand the windmill in a symbolic sense, we, as a people, wouldn't have taken thousands and thousands of pictures of it. And if we didn't, this book would not exist.

## Notes

1. *The (Valentine, Nebr.) Republican*, November 10, 1944, 1.
2. http://memory.loc.gov.

# CHAPTER I

# American Windmills

The windmill is a fondly recognized feature of the American landscape, a sentinel rising above rooftops and fields. Its stalwart presence states clearly that human ingenuity has been at work. From the earliest days of white settlement on the Atlantic coast, immigrants built windmills for grinding grain using Old World technology. By the 1850s, American inventors had developed smaller wind machines that appealed to farmers and livestock raisers. An entire industry developed that mass-produced windmills by the hundreds of thousands. These machines eventually appeared in all parts of America and were exported to markets around the world.

During the twentieth century, newer machines joined the ranks of those producing power by mechanical means. These generated the modern miracle —electricity—for homes, shops, and streets. Today, both windmills and wind generators remain on the market, filling a niche of the economy that uses free energy in areas electrical power grids do not reach or for consumers who prefer to use natural sources of power.

From the nineteenth century onward, the sculptural quality of windmills intrigued photographers. Sometimes windmills were the only subjects in

In 1886, members of the McCartney family of Dale Valley, Nebraska, posed with their Halladay Standard windmill, sod house, and animals for Solomon D. Butcher's camera in a classic frontier image. Courtesy of the Nebraska State Historical Society, Lincoln, Nebraska.

these images, though more often windmills were used as part of the general composition of a broad range of photographs. These are often the more interesting images, as they show the context for the wind machines in relation to everyday life in past decades.

Among the most appealing of all the images of windmills are the nearly five thousand glass plate negatives taken by Nebraska photographer Solomon D. Butcher that date from the 1880s to the 1900s. He consciously used windmills as landscape elements as he composed his photographs of settlers on the sod house frontier. The families and individuals who paid Butcher to make these photographs brought their most treasured household possessions outside, lined up their most handsome draft animals, and put on their best clothes for the photographer's camera. Because wells and windmills were among their most expensive improvements to the land, farmers and livestock raisers wanted them included in the pictures.

A new phase of photography began with the invention of the inexpensive handheld cameras that entered the consumer economy in the late nineteenth century. They made it possible for windmill owners to make their own photographs, resulting in thousands of historic photographs showing windmills from the perspective of ordinary people. Groups of boys clambered

*Built in the 1790s, this European-style windmill still stands in Chatham, Massachusetts. When a photographer took this picture in the early twentieth century, the canvas sails had been removed from its four wooden framework blades. From the author's collection.*

*The cover article in the October 7, 1854, Scientific American that described Daniel Halladay's new self-governing windmill. From the author's collection.*

*Daniel Halladay, who invented the first commercially successful self-governing windmill in the United States in 1854. Courtesy of the Batavia Historical Society, Batavia, Illinois.*

onto towers while friends snapped their pictures. Couples on courting excursions in horse-drawn top buggies and in motorcars stopped at windmills for water, picnics, and snapshots. Windmills stood in the background of pictures at weddings and reunions. These images form part of the record of life in the United States.

## Windmill History

Windmills first appeared in the present-day United States in the 1620s. English immigrants erected them first in Virginia and then in multiple locations along the Atlantic coast as well as inland. French-speaking immigrants built windmills in what became Canada. These traditional Old World windmills usually had four blades covered with canvas sails and had to be directed by hand to face the wind.

*A crew from a windmill distributor delivered this Challenge 27 mill, which was twenty feet in diameter, to a New Mexico ranch around 1935. It was made by the Challenge Company of Batavia, Illinois. Photo by and courtesy of B. H. Burdick, Sr.*

European-style windmills worked satisfactorily for the purpose of grinding grain for communities, but they were large, expensive structures. Most farmers or livestock raisers could never dream of owning such machines. They needed small wind machines for pumping water and grinding limited amounts of grain. As early as 1837, James Kerr of Maury County, Tennessee, patented an early wind pump for small-scale use, a design that apparently was not successful. What was needed was a windmill that was self-governing, that would automatically swivel to face the wind and control its speed of operation without human attention.

In 1854, a New England Yankee, Daniel Halladay, came up with the first design for a self-governing American windmill that was commercially successful. He and other capitalists organized a company in South Coventry, Connecticut, to produce the wind machines. The company promoted its windmills at state and county fairs and in periodicals for progressive farmers. Their Chicago agent tapped into a substantial market on the prairies of the Midwest and on the eastern fringe of the Great Plains. The expense and

*Factory-made windmills dotted farms throughout the United States. Here a Perkins Steel windmill made by the Perkins Wind Mill Company of Mishawaka, Indiana, pumped groundwater for a prosperous Midwestern farm in the early twentieth century. From the author's collection.*

delays of shipping windmills from Connecticut to Illinois led the company to move its manufacturing operation in 1863 to Batavia, Illinois, in the Fox River Valley just west of Chicago. There it operated as the U.S. Wind Engine and Pump Company from the 1860s into the 1940s.

Other entrepreneurs on the plains and prairies saw the market for windmills, and starting in the 1870s a substantial number of them opened factories. Consumers in the East also recognized the advantages of windmills, and several firms began to produce them in eastern states, as did others on the Pacific coast. Through the 1880s to the 1900s, windmills were a substantial part of the farm implement industry in the United States. Trade journals like *Farm Implement News* and *Implement Age* carried regular articles on windmill manufacturers and their products, and farm newspapers like *American Agriculturist* and *Hoard's Dairyman* carried advertisements aimed at potential buyers.

As more factory-made windmills appeared on the market, demand grew for people who could erect and maintain them. Many consumers assembled

PHOTO BY ROXELL — TOWN PUMP MAIN ST POLK OHIO,

*An Enterprise windmill mounted on an enclosed tower pumped water from the public well on Main Street in Polk, Ohio, in the early twentieth century. The Enterprise was manufactured by the Sandwich Enterprise Company of Sandwich, Illinois. From the author's collection.*

OPPOSITE:

*When crews brought electrical power to towns, consumers often switched from their backyard windmills to electric power pumps to supply their water needs. From the author's collection.*

and erected their windmills on individually built towers, but soon dealers, well drillers, and others began offering this service. In time, a group known as windmillers began to specialize in these services. On large ranches in the American West, owners employed crews specifically to make the rounds of their windmills to lubricate them regularly or make repairs when needed. The work of windmillers is still important today. In many rural areas, they rank among the most indispensable members of the community, together with preachers, teachers, and veterinarians.

Probably due to the influence of western novels and motion pictures, windmills are associated with ranching. This is valid because windmills made it possible for humans and their animals to live in the arid West. Properties like the storied XIT Ranch in the Texas Panhandle used windmills by the hundreds.

Ranchers bought many windmills, but farmers were the main customer base for windmill manufacturers. All over the United States, but especially

in the West and Midwest, agriculturists used the power of the wind to bring up groundwater that otherwise would have had to be pumped by hand. Many windmills were used in areas with comparatively dense rural populations, such as Illinois and the lower peninsula of Michigan. Almost every small farm that did not have naturally running water had at least one well with a windmill.

A fad for irrigation using windmills attracted agriculturists in the 1890s. In places like the Platte River Valley of Nebraska, the Arkansas River Valley of Kansas, and the San Joaquin Valley of California, many farmers invested substantial sums to sink wells, erect heavy-duty windmills, and construct large earthen water reservoirs to irrigate agricultural crops. Soon dams and canals for surface water and power pumps for groundwater supplanted windmills as the technology that made irrigation possible, but the historic earthen reservoirs still survive in some areas.

Railways also used windmills. They required reliable supplies of clean boiler water for their steam locomotives. In localities without surface water, companies sank wells and installed windmills, some of which were huge, over the wells. The wind machines pumped water into large elevated tanks fitted with spouts; these tanks could quickly fill locomotive tenders with fresh water. These wind-powered water systems dotted the landscape of the United States until diesel locomotives replaced steam engines in the mid-twentieth century.

Windmills were used in urban areas as well. Individuals and businesses drilled wells and installed windmills to provide for their water needs. In some instances, windmills pumped groundwater into reservoirs, providing the water supply for entire towns. Public wells and windmills, which were often located in the center of a town's commercial district, provided free water for people and draft animals. In the East, windmills sometimes were mounted on the tops of multistory buildings to pump water into wooden reservoirs; the water flowed by gravity to users inside and for fire protection. Resorts, which brought the luxuries of urban life to rural areas with scenic beauty, often used windmills to pump fresh groundwater for guests and employees.

Not all windmills came from factories. Some people built their own wind machines. They used recycled materials like pieces of obsolete farm machinery and scrap lumber, gaining in economy of construction what was often lost in efficiency. Many homemade windmills lacked the self-governing features of mass-produced machines and they frequently tore to pieces during high winds. Even so, inventive individuals enjoyed building and profiting from the use of their homemade wind machines.

The market for windmills in the United States began to soften in the 1890s with the introduction and increasingly widespread use of small internal combustion engines. Windmill production continued on a large scale, but the small engines (which were typically called gas engines but actually used a variety of fuels) enabled consumers to choose machines that were not dependent on the fickle wind. Farmers and ranchers could pump water, grind grain, or cut firewood anytime they chose with an internal combustion engine that burned a fossil fuel like gasoline or kerosene. The demand for windmills changed more dramatically beginning in 1935, when the Rural Electrification Administration began introducing inexpensive electrical power to the countryside. With electricity, farmers and ranchers connected to the "high line" could do innumerable tasks cheaply and conveniently. Yet even in the context of the new technology, windmills persisted in the countryside; those who lived far from power lines often used small wind generators to produce at least enough electricity to illuminate a few light bulbs and operate radios for home entertainment.

During the economic depression of the 1930s, the windmill manufacturing industry entered a decline from which it never recovered. Rationing of strategic materials during World War II made it difficult for most makers to produce windmills and most of them shifted to contracts to produce war matériel. The factories in the Fox River Valley of Illinois, which already had the machinery to shape brass into pumping cylinders, for example, began

Dempster Mill Manufacturing Company
employees in 1947 received raw materials
by rail for the windmills, pumps, and farm
implements they made in this Beatrice,
Nebraska, factory. This firm was one of the
handful of companies that continued to
make windmills into recent times. From the
author's collection.

to produce brass-cased artillery shells. After the war the market for windmills returned, but it never reached prewar levels. Company after company changed to other products, became wholesalers for water supply goods, or simply went out of business. A handful of the old companies held on to the trade that remained, and several survive to this day. These and a few new firms and importers continue to supply new windmills to buyers who live far from electric power or who still find it profitable or desirable to use the free power of the wind instead of costly fossil fuels.

Photographs document all of the many uses of windmills in the United States. They help us understand that consumers viewed windmills as utilitarian devices that could be put to work in almost any setting, whether on a mountaintop or in the heart of a city. These images also show us the many ways that inventive individuals designed and built their own homemade wind machines, preferring their own ingenuity to that of trained engineers in factories. They also demonstrate how users in other countries modified technology developed in America to meet their own needs in places as diverse as soft, green England and the harsh, arid Sahara. To this day, windmills attract photographers, both professional and amateur.

# CHAPTER 2

# Windmill Manufacture and Distribution

**W**indmill manufacture began in small metal- and woodworking shops, many of which were single rooms. The pioneer makers contracted with foundries to produce iron castings that they then smoothed and machined, adding other wooden and smaller metal pieces to produce the complete machines.

As the volume of production increased, successful manufacturers expanded into factories with multiple departments, producing windmills with interchangeable parts for mass markets. The typical windmill factory included offices, a pattern-making shop, a foundry, a machine shop, a woodworking department, a sheet metal department, an assembly shop, and a shipping department.

The work force in windmill factories consisted predominately of men. A small number of women found positions as stenographers and bookkeepers in the offices. Massive immigration from Europe in the late nineteenth century helped provide the workforce for such factories. In the Fox River Valley west

Workers standing outside the factory of the Plymouth Iron Wind Mill Company in Plymouth, Michigan, in 1889. Inventor Clarence J. Hamilton developed not only the windmills made by the firm but also Daisy air rifles, which the company reputedly gave away as premiums to customers who purchased windmills. Courtesy of the Plymouth Historical Society, Plymouth, Michigan.

Most large windmill factories had their own foundries where they produced iron castings. Here workers stand behind sand molds and a bucket for pouring molten iron in the foundry of the Challenge Wind Mill and Feed Mill Company in Batavia, Illinois, at the turn of the twentieth century. The stone walls and dirt floor helped reduce fire danger in this area. The main casting for a Single Header windmill leans against a mold box in the right foreground. Courtesy of the Batavia Historical Society, Batavia, Illinois.

Smaller manufacturers like the Kregel Windmill Company contracted out to foundries to produce castings. The Kregel employees in this photograph machined main castings and wheel hubs for the Eli windmills made in its factory in Nebraska City, Nebraska. Leather belts connected to steel shafts driven by an internal combustion engine powered all of the machinery here, which included drill presses, a large shearer, and a band saw. Courtesy of the Kregel Windmill Company Museum, Nebraska City, Nebraska.

This man spent each day operating a turret lathe in the factory of the Dempster Mill Manufacturing Company in Beatrice, Nebraska, during the 1950s. In the trough below the work surface and on the floor are the metal shavings created as he shaped precise metal parts. From the author's collection.

In the 1930s, George Walstrom assembled heads for Wonder Model B windmills at the factory of the Elgin Windmill Company in Elgin, Illinois. The head, which consists of the main casting and attached parts, is the mechanical portion of a windmill that converts the rotary motion of the wheel into an up-and-down reciprocating pump stroke. Courtesy of George A. Carlson.

of Chicago, for example, many workers were of Scandinavian origin; a company in this area named one of its windmills the Terrible Swede. Photographs of workers in windmill plants often suggest national origins. Their boyish faces show how early many started lives of labor in past decades.

Every windmill manufacturer either had or aspired to have its own foundry, where furnaces heated metal to liquid form, which was then poured into molds to make cast parts on the spot. A foundry enabled a company to produce its own iron castings when needed and to make repair parts when machines broke down. To produce castings, workers first made full-size wooden patterns of the parts required. They used these patterns to make molds in wooden frames containing special casting sand. Molders poured molten iron into the molds. After the metal cooled, foundry workers passed the iron parts to others to smooth and prepare for actual use. If the company could not afford its own foundry, it paid other firms to produce castings under contract arrangements.

Machinists used special machines to grind away imperfections on rough iron castings. Excess metal often flowed into cracks in the molds or created other rough edges on the parts. In larger factories, men took the smaller iron parts to an area known variously as the mill room, the tumbling room, or the rattle room. There they put the loose castings into rotating tumblers with iron balls and wear plates (special rubbing surfaces inside the tumblers).

As the cylindrical devices turned, imperfections, burrs, and excess metal wore away. All the detached iron went back into the furnace to be melted and used again. In smaller factories without tumbling machines, the rough iron castings went directly to the machine shop.

The machine shop was a key department in every windmill factory. There well-paid skilled workers used special grinders to remove imperfections from castings, smoothed cast-iron gears on hobber machines, cut steel gears, milled shafts, cut keyways, prepared bearings, and did most of the precise metalwork in the plants. Many larger factories also had separate blacksmith shops that provided maintenance and repair services for the entire facility.

In the early days of windmill manufacture, the woodworking department played a major role in production. Workers converted large wooden timbers into windmill wheels and vanes and produced prefabricated wooden towers and tanks. Every tool in the woodworking shop was designed for cutting, making it a dangerous place to work. The sawed and assembled wooden parts were next sent to the paint room, where most were dipped to give the product a finished appearance and a measure of protection from the elements.

In time, the demand for wooden windmills decreased as all-metal machines took their place and the sheet metal department supplanted the woodworking shop in importance. In this newer area, highly skilled workers cut galvanized sheet steel and shaped it into blades, vane sheets, and hoods. They also cut angle steel into vane stems and wheel arms and cut and punched angle steel (steel with lengthwise bends to promote rigidity) and other metal material to create prefabricated towers. Some larger companies had their own galvanizing plants to give steel parts zinc weatherproofing; smaller firms used steel that had already been coated with molten zinc.

Parts converged from the various areas of the factory in the assembly and shipping departments. There the elements of windmills, towers, tanks, pumps, and other pieces of associated water supply equipment came together. Workers crated and bundled the parts and pieces for shipment. As orders came in from distributors, goods were wheeled onto wooden or concrete platforms and loaded into boxcars for shipment by rail to customers.

Factory workers assembled this Monitor Vaneless Style L windmill outside the brick walls of the Baker Manufacturing Company in Evansville, Wisconsin, about 1920 for a series of publicity photographs, including this close-up view. From the author's collection.

Company workers erected this tall wooden tower to test an Industrious windmill in the factory yard outside the Avery Planter Company in Peoria, Illinois, around 1890. The photograph, possibly made from the factory building, shows the entire area covered with a light dusting of snow. From the author's collection.

After they assembled and crated windmills
and related products, workers used wheeled
dollies to roll them up ramps onto loading
platforms and then into boxcars for
shipment to wholesalers by rail. Courtesy
of the Baker Manufacturing Company,
Evansville, Wisconsin.

# INSIDE THE CHALLENGE WINDMILL FACTORY

During the second half of the nineteenth century, Batavia, Illinois, in the Fox River Valley became one of the most important centers of windmill manufacture in the United States. Today, the town boasts multiple historic factory buildings where workers built hundreds of thousands of windmills.

As early as 1863, the manufacture of Halladay's patent windmills shifted from Connecticut to Batavia. These were the first commercially successful self-governing American windmills, invented in 1854 by Daniel Halladay. Other important windmill makers followed him to Batavia, including the Challenge Wind Mill and Feed Mill Company and the Appleton Manufacturing Company. They joined Halladay's firm, the U.S. Wind Engine and Pump Company. All three sold to national markets. Drawn by the convenient shipping of the region and access to water power for their factories, a number of smaller windmill makers sprang up from time to time in Batavia as well.

Many photographs of local windmill factories and workers were made in Batavia. The Batavia Historical Society holds a treasure trove of historic photographs showing all aspects of windmill manufacture. The images that illustrate the production methods at the Challenge Wind Mill and Feed Mill Company around the turn of the twentieth century are unquestionably the best.

*Men and women who worked in the spacious second-floor offices of the Challenge Wind Mill and Feed Mill Company around 1900. This prosperous company provided steam heat and electric lights for its white-collar staff. Courtesy of the Batavia Historical Society, Batavia, Illinois.*

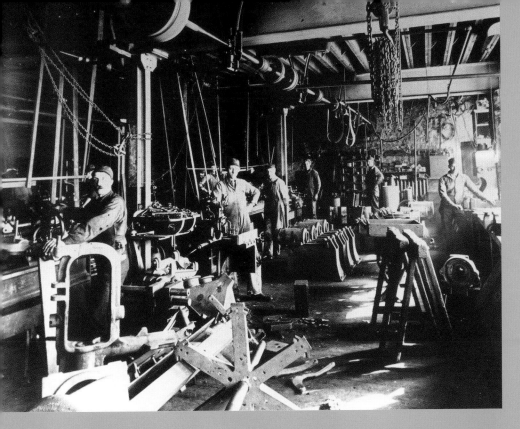

The machine shop in the plant of the Challenge Wind Mill and Feed Mill Company around the turn of the twentieth century. Here workers using machines powered by leather belts from spinning overhead shafts shaped iron and steel parts for a wide range of windmills and pumps. The man on the left is resting his hand on the main casting for a Dandy Irrigator windmill. Courtesy of the Batavia Historical Society, Batavia, Illinois.

The woodworking department at the Challenge Wind Mill and Feed Mill Company at the turn of the twentieth century. Here skilled workmen converted rough lumber into windmill blades, arms, vanes, and other components. Courtesy of the Batavia Historical Society, Batavia, Illinois.

# CHAPTER 3

# Windmill Marketing

**E**ven before factory workers loaded new windmills onto railway cars for shipment, the work of marketing had preceded them. Some of the best minds in the industry focused on sales.

The front line in marketing was traveling salesmen. These "traveling men," as they were called, went from town to town, where they visited farm implement dealerships, hardware stores, and lumberyards; distributed sales literature; and promoted contracts for the companies they represented. Sales staff assembled exhibits at county and state fairs, where they erected tents to shelter and welcome customers while pitching the merits of their particular lines.

World's fairs in the late nineteenth and early twentieth centuries were the sites of the most spectacular commercial exhibits of windmills. The largest of these displays sprang up beside a lagoon on the grounds of the World's Columbian Exposition at Chicago in 1893. In its October 12, 1893, issue, *Farm Implement News* described the windmills at the exposition: "Numerous tall towers, like the masts of ships grouped closely in harbor with colors flying, surmounted by wind wheels of various forms and sizes whirling in the bright sunlight, and throwing off sparkling rays of many colors, afford from the distance an unique, lively and brilliant spectacle." This description was no exaggeration.

Printed advertisements and trade literature were the most common forms of promotion. Manufacturers targeted some advertisements toward individual consumers, but they directed most of their efforts toward potential dealers. Company participation in conventions and other gatherings of dealers became a necessary element of sales promotion. Such meetings gave sales staff the chance to meet personally with dealers and put sales literature and catalogs directly into their hands.

*Many retail dealers went to great lengths to promote sales of their goods, as did this Aermotor Company agent in Canada. Despite a dusting of snow, he created this huge display of vane sheets and crated wheel sections for self-lubricating steel mills about 1915. In the background is a multistory wooden grain storage elevator that dominated the prairie town. From the author's collection.*

Many farm implement dealers erected windmill displays at county fairs in the late nineteenth and early twentieth centuries. In this instance, a dealer erected an Iron Turbine windmill made by Mast, Foos and Company of Springfield, Ohio, amid the rows of plows, cultivators, and other equipment at a New York State county fair about 1890. From the author's collection.

The Fort Wayne Wind Mill Company of Fort Wayne, Indiana, created this commercial exhibit at an unidentified fair sometime during the 1890s. From the author's collection.

Once windmill makers created wholesale business relationships with distributors, the agreements might last for decades. The Axtell Company in Fort Worth, Texas, for example, sold windmills, towers, and pumps made by the Baker Manufacturing Company in Wisconsin from the 1890s to the 1950s, some sixty years. For most of this time, the Axtell Company enjoyed exclusive distribution rights to the Monitor line of windmills in both Texas and Oklahoma. This meant that the Baker firm wholesaled its water supply goods only to the Axtell Company. Axtell then jobbed the Monitor windmills, pumps, and accessories to individual retailers.

Some companies adopted different strategies to promote their goods. Although most manufacturing firms sold windmills, towers, pumps, and auxiliary goods only to dealers, a handful of firms sold directly to customers. These were mostly smaller firms like the Kregel Windmill Company of Nebraska City, Nebraska, and the Ottawa Manufacturing Company of Ottawa, Kansas. They sold windmills to buyers within about fifty miles of their plants, near enough for farmers to pick them up in wagons or early motorcars.

A number of mail-order firms also distributed windmills. They contracted with windmill manufacturers to produce mills marked with their own brand names. In this way Sears, Roebuck and Co. sold Kenwood windmills for decades, while Montgomery Ward & Company distributed Clipper mills. Some smaller mail-order catalog companies also had their own windmill lines.

Corinne Baker dressed the part to promote Perkins windmills sold by C. Titus in San Angelo, Texas, in the late nineteenth century. Her headdress supports a miniature Perkins Solid Wheel windmill, while her skirt simulates the pattern of wooden blades in a wheel. Courtesy of Fort Concho National Historic Landmark, San Angelo, Texas.

Using an electrically lighted sign to attract prospective customers and a marquee tent to shelter them from the sun and from inclement weather, these salesmen at an unidentified state fair in the 1920s promoted sales of Monitor Vaneless and Monitor Steel windmills made by the Baker Manufacturing Company of Evansville, Wisconsin. Courtesy of the Baker Manufacturing Company, Evansville, Wisconsin.

When multiple distributors erected windmills at county fairs like this one in Garden City, Kansas, in 1898, customers could compare and contrast the performance of competing brands. Courtesy of the Kansas State Historical Society, Topeka, Kansas.

The commercial windmill exhibits at the 1893 World's Columbian Exposition in Chicago as viewed behind the pavilions for French colonies in Southeast Asia and North Africa. These major world's fairs gave people from around the world opportunities to demonstrate their national pride. Courtesy of the Chicago Historical Society, Chicago, Illinois.

On the flat prairie of the southern Great Plains, employees of the Western Windmill Company put together this animal-drawn float for the Independence Day parade in Lubbock, Texas, in 1911. To illustrate the happy home life made possible by windmills, the float carried a cookstove, trunks, and pots and pans—and a miniature Samson windmill. Courtesy of the Southwest Collection, Texas Tech University, Lubbock, Texas.

This just-returned American doughboy still in his World War I-era uniform used a Model T Ford to go to the factory to pick up an Eli windmill made by the Kregel Windmill Company in Nebraska City, Nebraska, in the late 1910s. Courtesy of the Kregel Windmill Company Museum, Nebraska City, Nebraska.

The Axtell Company of Fort Worth, Texas, celebrated its golden anniversary in 1941 with this custom trailer built to advertise the windmills and other water-supply goods it distributed. Courtesy of the Baker Manufacturing Company, Evansville, Wisconsin.

After World War II, many windmill manufacturers shifted part of their marketing efforts to regional and national conventions for drillers of water wells and distributors of water-supply equipment. Sales staff from the Aermotor Company of Chicago assembled this display for the Oklahoma Hardware Show in February 1952. From the author's collection.

C. J. Goulet and Burwell N. Kilbourn (left and right front) of the Dempster Mill Manufacturing Company of Beatrice, Nebraska, enjoy cocktails with colleagues and customers at an Omaha convention in 1954. From the author's collection.

B. H. Burdick, Sr., crouched at the right in this early 1950s photo, purchased this Stinson Station Wagon airplane in 1946 to advertise his business that wholesaled windmills and other water-supply goods in El Paso, Texas. Courtesy of B. H. Burdick, Sr.

# MAIL-ORDER WINDMILLS FROM SEARS

*Three men on the Great Plains stand beside a Model T Ford touring car parked in front of a Kenwood steel windmill distributed by Sears, Roebuck and Co. At the base of the tower behind them is a circular steel tank for watering livestock. From the author's collection.*

Sears, Roebuck and Co. was one of several companies that sold windmills through their mail-order catalogs. The Chicago-based firm, established in 1886, began distributing windmills made under contract by other firms in 1896. Shortly thereafter, Sears adopted the name Kenwood for its windmill product line, which it sold to buyers in all parts of the country.

During the early twentieth century, Sears, Roebuck and Co. adopted a novel approach to promoting its mail-order windmills. It offered customers the opportunity, at no cost, to personalize the vane sheets of their mills. The buyer could choose to have the Kenwood brand name on its tail, to have nothing whatever painted there, or to have his or her own name painted on its sides. Sears explained to its customers, "Realizing that many of our customers would like to have their name on the rudder of their windmill, and with a desire to accommodate our customers in every possible manner, we have arranged so that when they wish to have this done we will stencil their initials and their surname on both sides of the rudder sheer. . . . We cannot stencil more than the initials and the name . . . because there is not room for more." The catalog directed buyers to write on the order form that they wanted a name added and to print the requested name on a slip of paper clipped to the order.

This sales promotion resulted in Sears windmills across the country that bore their owners' names. Preserved in historic photographs, these names perpetuate the memories of the original buyers of the mills.

This ten-foot-diameter Kenwood solid-wheel wooden windmill and its drilled well, photographed at the headquarters of a ranch near Buckingham, Colorado, around 1900, probably cost more than the tar-paper shack with the stone base around its walls. Because Sears offered its customers the option of having their names painted on the vanes of their windmills, we know that this was the ranch of Ed Skanes. From the author's collection.

A woman photographer identified as Mrs. Vierte recorded two little girls standing on the gallery porch railings of the A. B. Overman home in Crookston, Nebraska, in the early twentieth century. The Kenwood steel windmill behind the house, sold by Sears, Roebuck and Co., bears the stenciled name of the homeowner on its vane. From the author's collection.

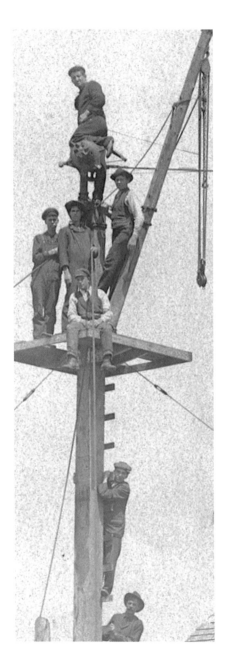

# CHAPTER 4

# Windmill Erection and Maintenance

**A**s windmills became a part of life during the latter half of the nineteenth century, a demand arose for service people to erect and care for them. Often dealers erected the mills themselves or contracted the job out. Often the job went to well drillers, many of whom simply added the erection and care of windmills to their regular tasks of sinking wells and repairing pumps. On the big ranches in the American West, owners delegated responsibility for maintaining sometimes dozens of windmills to employees. These men spent their work hours riding across the expanses of pasture, going from windmill to windmill, endlessly greasing and often repairing them. These workers came to be known as windmillers.

Farmers typically took care of their own windmills unless the repairs were too complex. Almost all the older wooden-wheel and open back-geared steel mills required regular lubrication, often weekly. It was not unusual for the sons in families to be assigned this duty as soon as they were old enough to climb the towers. This work presented no great difficulties during pleasant weather but became hazardous during high winds or when ice and snow covered the ladders and service platforms.

Repairs might be as simple as tightening loose bolts. Other times, however, workers replaced worn bearings or gears, repaired wind-damaged wheels and vanes, and even replaced broken main shafts. Frequently maintenance had to be done on the piston displacement pumps, which were located underground at the water table. These pumps, which were actuated

Before a windmill could be erected, a well had to be drilled or dug by hand. Generally the wells for windmills were drilled, as they were both faster and cheaper than hand-dug wells at most depths. Here photographer Solomon D. Butcher recorded work by a crew from Hart and Company using a horse-powered machine to operate a percussion drilling rig on the treeless plains at Cliff Table, Nebraska, in 1890. The horses walked in a circle around a mechanism that connected by way of gears and a shaft to the drilling rig. The rig repeatedly raised a heavy metal drill bit and then let it drop to peck a hole in the ground one blow at a time. Courtesy of the Nebraska State Historical Society, Lincoln, Nebraska.

After drilling a well, this unidentified crew installed a ten-foot-diameter Original Star windmill made by the Flint and Walling Manufacturing Company of Kendallville, Indiana. The men had hitched mules to their drilling rig and were about to leave the job site when someone snapped this photograph around the turn of the twentieth century. From the author's collection.

by the wooden pump rods that extended down from the windmill, pushed the water up the pipes to the surface. When their leather seals became worn, whoever was doing the repairs had to lift out the pumps to replace the gaskets.

Some of the heaviest work for windmillers involved erecting new machines. Laborers began these jobs by hand-excavating anchor holes in the ground to secure the legs of the tower. The minimum depth for these anchors was five feet, every inch of which was dug with picks and shovels. The laborers then erected the tower, using one of two methods. When it was practicable, the windmill erectors assembled the tower on the ground and used cables drawn by animals, motor vehicles, or hand winches to raise the derrick into a vertical position over the well. When space was limited or the unusual height of a tower precluded this method, crews built the tower up one piece at a time. In this method, workers used gin poles to lift the windmill components to the top of the tower in order to put them together.

*Windmill work was done in fair weather and foul. Here two men have prepared to raise the tower for a Challenge Steel windmill made by the Challenge Company of Batavia, Illinois, in blowing snow sometime in the early twentieth century. From the author's collection.*

*There are two ways to erect windmill towers. One is to assemble the tower on the ground and then raise it with cables to a vertical position. The other way is to put the tower together one piece at a time to build it up from the ground. Here two men pose with a wooden tower they have constructed on its side just before raising it over a well near Carrizozo, New Mexico, about 1935. Photo by and courtesy of B. H. Burdick, Sr.*

Over time, the number of trained windmillers began to decline. To deal with this situation, schools such as New Mexico State University in Las Cruces, New Mexico, and Angelo State University in San Angelo, Texas, began offering short courses in windmill technology in the late 1970s and early 1980s to train new generations of erectors and repairers. Today the American Wind Power Center in Lubbock, Texas, continues to offer these training courses for those who would like to learn how to provide practical windmill service.

An unidentified Mexican worker stands beside the wheel of an eight-foot Challenge 27 windmill as a crew from the Burdick & Burdick Company raises it for Dr. J. D. Love at Hatch, New Mexico, about 1935. The square wooden platform on the tower is there to provide footing for service and repairs. Photo by and courtesy of B. H. Burdick, Sr.

Ropes were one of the keys to successful windmill work during the late nineteenth and early twentieth centuries. Here a windmiller spreads the weight of his ropes across his shoulders in order to carry them more easily. From the author's collection.

Work on top of windmills was never employment for the faint-hearted. About 1935, B. H. Burdick, Sr., took his camera to the apex of a windmill tower, pointed it down toward the ground, and snapped this picture. His feet straddled the head of a Challenge 27 windmill, his shoes and a pair of pliers giving scale to a remarkable view. Photo by and courtesy of B. H. Burdick, Sr.

OPPOSITE:

After a crew from the Burdick & Burdick Company built this fifty-foot steel tower up from the ground one piece at a time, they used a gin pole with blocks and tackle to lift the head of a huge twenty-two-foot Challenge 27 windmill to its place atop the tower in the desert Southwest in the 1940s. The circular stone tank beside the mill held water. Photo by and courtesy of B. H. Burdick, Sr.

Windmillers in Canada used this gin pole to lift components of a Canadian Steel Airmotor made by the Ontario Wind Engine and Pump Company of Toronto to position on a mast-style tower above a barn in the early twentieth century. Windmill crew members generally liked to pose for pictures of work atop towers. From the author's collection.

During the twentieth century, motor vehicles took the place of horses and wagons to deliver windmills to job sites. Here a Burdick & Burdick Company employee stands in the back of a GMC truck outside the company warehouse in El Paso, Texas, around 1941. The vehicle is loaded with two Challenge 27 windmills and their towers. Photo by and courtesy of B. H. Burdick, Sr.

For small jobs, a passenger car with a trailer could carry everything needed for two men to erect a new windmill and tower. Burdick & Burdick Company employee Carl Boyd smiled for a photographer as he reclined on a pile of gear on such a trailer after finishing a job in the desert Southwest in 1942. Photo by and courtesy of B. H. Burdick, Sr.

OPPOSITE:

A man inspecting a section of steel blades from a sixteen-foot Challenge 27 windmill in the desert Southwest about 1935. In the background, a Steel Eclipse Type WG windmill made by Fairbanks, Morse and Company of Chicago pumps water. Photo by and courtesy of B. H. Burdick, Sr.

An unidentified group of windmillers,
including an African American worker, rests
beside a campfire in the Southwest during
the mid-1930s. Windmill crews carried
enough sleeping and cooking gear to spend
several days at a time in remote locations.
This group has nailed a wooden packing
crate onto a tree to create their own
"kitchen cabinet" for groceries. Photo by and
courtesy of B. H. Burdick, Sr.

OPPOSITE:

A shortage developed in the 1970s for
technicians who were trained to install
and repair windmills. New Mexico State
University (Las Cruces, New Mexico) and
Angelo State University (San Angelo, Texas)
both began offering weeklong short courses
in windmill technology to train well service
personnel and others the traditional skills of
windmillers. Here students erect a new
windmill on the campus of Angelo State
University in the early 1980s. From the
author's collection.

# WINDMILL WORK AND PEARL HARBOR

B. H. "Tex" Burdick, Sr., of El Paso, Texas, was one of the most famous of the old-time windmill men. People today are familiar with Burdick from the collection of photographs he made from the 1920s to the 1940s that depict his employees doing all types of windmill work. Burdick preserved his file of several hundred photographs and negatives until the 1980s, when he allowed the author to copy them and make them available to the public at the Panhandle-Plains Historical Museum in Canyon, Texas.

Among B. H. "Tex" Burdick's photographs are a series made in December 1941. The occasion was the erection of an impressive eighty-foot combined tank and windmill tower on the Warner Ranch north of Rodeo, New Mexico. This was a big job, and Burdick himself assisted three of his best men in the undertaking. After hand-digging four extra-deep anchor holes, the men built the tower up from the ground, one piece at a time. When they reached sixty feet (the height of a six-story building), they put together a big water tank made of wood staves, placing it inside the tower. The water the windmill pumped would flow by gravity from this tank to the home, barns, and corrals at the ranch headquarters. After completing the wooden tank, the crew extended the tower another twenty feet into the air. The final stage consisted of assembling a ten-foot-diameter Challenge 27 windmill at the top.

The Warner Ranch job was memorable for Burdick and his men for more reasons than just its magnitude. Tex was suffering from a terrible cold, and his nose was running constantly. He later recalled that Kleenex paper tissues had recently come onto the market, and he used them liberally. The weather was cold and very windy. "About every two or three minutes I'd grab that Kleenex and then turn it loose, and it would go out there a mile and a half. . . . I'll never forget that." Windmillers' camps were not necessarily the tidiest of places.

*Tony Venagas, Harry Clifford, and Carl Boyd (from left), all windmillers with the Burdick & Burdick Company, chow down on breakfast at the Warner Ranch north of Rodeo, New Mexico, in December 1941. Photo by and courtesy of B. H. Burdick, Sr.*

The men had other memories of the Warner Ranch. While they were listening to the automobile radio on site, they learned of the Japanese attack on Pearl Harbor on December 7, 1941. The four men had retreated from the cold and wind to the automobile, where they listened late into the night to sketchy news reports of the sneak attack on the U.S. Pacific fleet in the distant Territory of Hawaii. "Then all of a sudden," Harry Clifford remembered, "about midnight these airplanes started coming." The men did not know where they had come from or where they were going, but the effect on them was electrifying. "We didn't know whether we ought to quit and go and sign up or wait till they came and got us." Burdick added, "I left the radio on in the car, and the next day we had to push it to get it started."

*B. H. Burdick, Sr., and his men built this combined windmill and tank tower sixty feet high and then assembled a tapered wooden water tank from that height. Photo by and courtesy of B. H. Burdick, Sr.*

*After finishing the eighty-foot tank tower and assembling a ten-foot-diameter Challenge 27 windmill at its apex, Harry Clifford stood on top while his boss, B. H. Burdick, Sr., snapped this picture. Courtesy of B. H. Burdick, Sr.*

# CHAPTER 5

# Windmills and Western Ranching

**W**hen many people think of windmills, the first image that comes to mind is of a single tower standing in a windswept cattle pasture. Pumping for livestock was one of the first uses of factory-made windmills in North America. This role of wind power is still important today, as there are still many areas far from electrical power where windmills by the thousands pump water for stock.

Windmills were instrumental in making ranching possible in the western areas of the United States and Canada. Wherever water could be found underground, wells and windmills could make it available for users.

This ability to have water where a rancher wanted it, or at least where he or she could afford it, meant that property could be divided into pastures with inexpensive barbed-wire fencing. For many livestock raisers, this combination of drilled wells, windmills for pumping, and wire fences allowed controlled breeding of their animals for the first time.

Initially ranchers tried to make water available for the cattle every five miles across their properties. As the marketing of cattle changed from a per-head to a per-pound basis, they began sinking more wells and putting up more windmills so their animals would not walk off weight in order to get to water.

A typical pasture water system consisted of a windmill over a drilled well, an open round metal or wooden reservoir with sides low enough to

*Two women and a man stop beneath a Standard windmill made by the F. W. Axtell Manufacturing Company, Fort Worth, Texas, during a horseback ride across a ranch on the southern Great Plains during the early twentieth century. From the author's collection.*

*This family posed for a picture by Solomon D. Butcher in front of their ranch head-quarters in Cherry County, Nebraska, in 1901. A Dempster No. 1 Vaneless windmill made by the Dempster Mill Manufacturing Company of Beatrice, Nebraska, pumped their water. Courtesy of the Nebraska State Historical Society, Lincoln, Nebraska.*

allow cattle to drink, and an overflow pipe to carry excess water to an earthen pond. Ranchers often ran their windmills twenty-four hours a day, turning them off only when necessary for service, pumping water as long as the wind blew. It was consumed by animals, it evaporated, or it sank into the ground at the overflow ponds called stock tanks.

Windmills were important landmarks on the plains. They sometimes even doubled as lighthouses; ranch wives occasionally climbed towers to hang kerosene lamps so husbands or other family members could find their way back home across the prairie after dark. Wherever travelers might go on the plains, they knew that the windmills dotting the horizon signaled places where water could be found.

This open back-geared Ideal Steel windmill made by the Stover Manufacturing Company of Freeport, Illinois, stood near a two-story stone ranch house on the Great Plains at the end of the nineteenth century. From the author's collection.

A Cyclone windmill made by the Pacific Pump and Wind Mill Company of San Francisco provided water in rectangular wooden troughs for horses at this ranch headquarters in California at the end of the nineteenth century. From the author's collection.

WINDMILLS AND WESTERN RANCHING   69

In 1886, Solomon D. Butcher made this photograph of members of the Stillman Gates family near the cattle pens on their ranch at Gates, Nebraska. In the background is an Iron Turbine windmill, the first commercially successful all-metal windmill sold in the United States. Note the sod roofs on some of the sod buildings. The Turbine was manufactured by Mast, Foos and Company of Springfield, Ohio. Courtesy of the Nebraska State Historical Society, Lincoln, Nebraska.

Members of this Cherry County, Nebraska, family stood in mud and slushy snow outside their substantial sod ranch house as Solomon D. Butcher made this photograph in 1901. Behind the house is a Dandy open back-geared steel windmill made by the Challenge Wind Mill and Feed Mill Company of Batavia, Illinois. Courtesy of the Nebraska State Historical Society, Lincoln, Nebraska.

A rancher in eastern New Mexico around 1900 took no chances when he found good water. He drilled a second well nearby, equipping both of them with Railroad Eclipse windmills made by Fairbanks, Morse and Company of Chicago. In the summertime, cattle soaked their feet in the overflow pond. From the author's collection.

The southern Great Plains and the
Southwest were strongholds of old-style
wooden-wheel windmills during the early
twentieth century because ranchers liked
their reliability and ease of repair. Some of
the examples were magnificent, like this
railroad-pattern Standard mill made by the
F. W. Axtell Manufacturing Company of Fort
Worth, Texas. Its diameter was twenty-two
and one-half feet. B. H. Burdick, Sr., recorded
it on film at the W. N. Fleck Ranch south of
Alamogordo, New Mexico, about 1923.
Courtesy of B. H. Burdick, Sr.

OPPOSITE:

Windmills enabled ranchers to occupy
remote desert locations, where sometimes
dozens of acres of sparse pasture were
required for each grazing animal. Here a
Turbine iron windmill made by the American
Well Works of Aurora, Illinois, pumped water
in a corral area near the Gila River in New
Mexico about 1920. From the author's
collection.

Solid-wheel vaneless wooden windmills were popular in California. These windmills did not have conventional vanes to direct their wheels into the wind. Instead, they had downwind wheels that turned behind the towers and swiveling vanes mounted parallel to the wheels. When an operator on the ground turned the mill on, the paddle-like vane pivoted to a horizontal position so the wheel would swing into the wind. When the operator shut the mill off, the paddle-shaped vane swiveled to a vertical position to push the wheel out of the wind. Here three men pose with a solid-wheel wooden vaneless windmill and a large wooden water tank on a California ranch at the turn of the twentieth century. From the author's collection.

Some ranchers used wind for more than pumping water. A livestock raiser in Canada erected this Power Aermotor windmill made by the Aermotor Company of Chicago atop his log barn to grind feed grains for his animals in the early twentieth century. From the author's collection.

Four African American children gathered for a photograph in front of an Eclipse windmill and an automobile in the ranch country south of Lubbock, Texas, in 1942. From the author's collection.

Windmills brought special luxuries to ranch families, among them the opportunity to swim in large stock tanks. The vane inscription on the open back-geared steel Samson windmill noted that it was distributed by Krakauer, Zork and Moye Hardware Company of El Paso, Texas, so this circa 1925 photograph must have been made somewhere in the desert Southwest. From the author's collection.

A regular-pattern Eclipse windmill pumped water for this home in ranch country on the Great Plains during the early twentieth century. Groundwater made the morning glories decorating its front possible. From the author's collection.

In a view that was repeated thousands of times on the treeless Great Plains of North America, a large oil-bath steel windmill pumped groundwater for cattle in western Nebraska about 1925. From the author's collection.

*Farm Security Administration photographer Dorothea Lange recorded this battered regular-pattern Standard windmill in the wind-blown sand of the Dust Bowl near Dalhart, Texas, in June 1938. Courtesy of the Library of Congress, Washington, D.C.*

# WINDMILLS ON THE XIT RANCH

In the early 1880s, officials of the state of Texas signed an unusual agreement with a group of Chicago investors. Texas, formerly an independent republic, entered the United States in 1845 with treaty provisions that allowed it to retain ownership of its public lands. For decades it had been land poor, owning millions of acres of vacant, nearly worthless property. The state needed a new capitol building, and the capitalists were interested in state-owned lands. The state agreed to give the investors, who came to be known as the Capitol Syndicate, 3 million acres of land if they would fund the construction of the new capitol building, and a deal was struck.

The businessmen funded most of the costs of constructing the capitol building, which is still in service in Austin, and in exchange they received a two-hundred-mile tract of land along the western side of the Texas Panhandle. They planned to sell the land to farmers, but they had to wait years before agricultural settlement reached that far west. To avoid losing money on the investment, they created what became the largest ranch under fence in the history of the United States, the XIT Ranch.

In order to stock the property with cattle successfully, the managers of the ranch cut it up into pastures using barbed-wire fencing and erected approximately 325 windmills to pump water from wells for both cattle and human use. Competing windmill makers struggled to secure the contracts to provide these mills, with most of the jobs going to agents of Fairbanks, Morse and Company, maker of the Eclipse windmills. The ranch was well known in its day for its hundreds of windmills.

*White-faced Hereford cattle graze in a pasture on the XIT Ranch where a regular-pattern Eclipse windmill pumped water in the early twentieth century. Courtesy of the Panhandle-Plains Historical Museum, Canyon, Texas.*

A ranch wife on the XIT Ranch feeds her flock of chickens about 1900. In the background a regular-pattern Eclipse windmill pumps water into an overhead storage tank. Courtesy of the Panhandle-Plains Historical Museum, Canyon, Texas.

This Railroad Eclipse windmill pumped groundwater to an elevated storage tank that supplied a dwelling on the XIT Ranch near Channing, Texas, in August 1902. Courtesy of the Panhandle-Plains Historical Museum, Canyon, Texas.

# WINDMILLS PUMPING OIL

Windmills were used to pump more than water. In at least one instance, American windmills were used on a Caribbean island to pump molasses. But the most common liquid other than water that windmills pumped was petroleum.

When modest oil deposits were not too deep, windmills could be adapted to elevate petroleum to the surface using the same type of pumps used for water. Throughout the American West, oil developers saw the potential for using the free power of the wind to elevate their "black gold."

Perhaps the best-known example of using wind power to pump oil in the twentieth century took place near Hobbs, New Mexico. There the mills elevated oil that had escaped from much deeper wells and had become trapped in the comparatively shallow water-bearing Ogallala Formation. In the mid-1960s, the Windmill Oil Company installed a battery of Aermotor windmills to pump oil from its subsurface location above the water in the aquifer. In the first three years, the mills extracted about 150,000 barrels even though state officials had said that it could not be recovered economically using traditional oil-field pump jacks. Developers used wind to pump oil in many locations throughout the American West.

*This regular-pattern Eclipse windmill pumped shallow oil into an open earthen tank near Toyah, Texas, in 1911. From the author's collection.*

In 1954, Hugh Burns Hutchinson repaired a collection of derelict windmills to pump shallow oil in the Conejo Oil Field in California. "BOCO" stood for Burns Oil Company. Photo by William T. Rintoul. Courtesy of Mrs. Frankie Jo Rintoul.

81

# CHAPTER 6

# Windmills on Farmsteads

**P**robably more windmills were used on farms than in any other setting. Thousands upon thousands of windmills left Midwestern factories for sale to farm families throughout the country. There they typically elevated groundwater for both humans and animals.

The elemental farm water system was a windmill on a tower standing over a drilled or dug well. In the simplest such systems, water from underground might pass through open troughs, with the overflow passing into earthen stock ponds. In slightly more sophisticated systems, windmills pumped water into elevated storage reservoirs from which the farm family could release the flow through faucets. Any overflow passed to open troughs and then into earthen tanks for animal use.

More sophisticated farm water systems used float valves connected to windmill regulators. These devices turned windmills off and on automatically when storage reservoirs filled or emptied. In colder parts of the country, some farmers placed elevated reservoirs in the attics of their homes or the lofts of their barns. In these locations, heat from the residents and the animals kept reservoirs and pipes from freezing in the wintertime.

Some agriculturists used windmills to irrigate their crops. They excavated earthen reservoirs, heaping the soil up at the edges, and allowed windmills to pump groundwater into them around the clock. When the time came to irrigate intensively cultivated fields of fruit or vegetables, they released the water to flow through open channels to the crops. There was a flurry of interest in windmill irrigation in the 1890s, but other methods supplanted the wind for irrigation in the decades that followed. Although windmills have not been used for serious crop watering for a century, the land still shows the marks of the circular and rectangular reservoirs that were constructed to hold irrigation water pumped by windmills.

*For a sodbuster, the well and windmill often were the most expensive capital investments when starting a farm. Members of a family stood in front of their handsome new ten-foot Original Star windmill made by the Flint and Walling Manufacturing Company of Kendallville, Indiana. They were photographed by Solomon D. Butcher at their simple sod house and storm cellar in Custer County, Nebraska, in 1887. Courtesy of the Nebraska State Historical Society, Lincoln, Nebraska.*

Schools used windmills in rural areas. Without water from other sources, schools often had their own wells and wind machines to pump water for pupils and teachers and the animals that transported them to the classrooms.

Windmills served a number of unexpected functions on farms. Owners often stocked windmill-filled ponds with desirable fish, which provided variety for the family dinner table. In northern states, these ponds formed layers of ice in the winter that some farmers cut and stored for use during the warmer months. Children enjoyed skating on the ice before it was cut. During the summer months, ponds were used for swimming and boating. Many people (including the author) learned to swim in such ponds filled with groundwater.

*Family members gathered in front of their sod house, all the draft animals drawn up into view, for this photograph by Solomon D. Butcher in 1889 or 1890. Behind the house, a Challenge Sectional Wheel windmill pumps water for this farm near Alger in northwestern Custer County, Nebraska. Courtesy of the Nebraska State Historical Society, Lincoln, Nebraska.*

*This farm family used a Joker windmill manufactured by a series of makers in Peabody, Kansas, from the 1880s to the 1900s. From the author's collection.*

*This man and woman posed for photographer Solomon D. Butcher in front of their Challenge Sectional Wheel windmill and round wooden stock-watering tank at the J. B. Davis farm near Ansley, Nebraska, in 1886. Courtesy of the Nebraska State Historical Society, Lincoln, Nebraska.*

A prosperous farmer somewhere in the
Northeast around the turn of the twentieth
century used an open back-geared
Pumping Aermotor windmill that was
mounted on a special patented tilting steel
tower. The hinged derrick permitted owners
to lower the mill for lubrication without
having to climb the tower. It was manufactured by the Aermotor Company of
Chicago. From the author's collection.

A ten-foot-diameter regular-pattern Eclipse windmill made by Fairbanks, Morse and Company of Chicago pumped water on this early-twentieth-century farm into a circular tank that provided drinking water for horses and other livestock. From the author's collection.

This well-to-do family gathered outside its two-story farmhouse in the vicinity of Upper Sandusky, Ohio, to show off not only children, toys, and fancy clothing but also a ten-foot-diameter Original Star windmill. Note the decorative brackets supporting the service platform on the wooden tower. From the author's collection.

A Power Aermotor made by the Aermotor Company of Chicago stood atop a wooden mast above this barn on a Midwestern farm in the early twentieth century. The farmer stored his summer hay in the barn for winter feeding and used the power windmill to grind grains to give his animals the highest nourishment. From the author's collection.

Cotton farmers, probably in Hemphill County, Texas, paused with their wagon beneath their regular-pattern Eclipse windmill about 1910. A trench for burying an underground water line extends from the base of the tower to the lower left corner of the picture. Courtesy of the Panhandle-Plains Historical Museum, Canyon, Texas.

An unidentified agriculturist on the Great Plains around the turn of the twentieth century brought all of his draft animals into the barnyard beneath a regular-pattern Eclipse windmill to document his economic success in this image taken by a professional photographer. From the author's collection.

This prosperous young couple, proud of their
two-story home and open back-geared
Samson steel windmill, posed for a picture
by Solomon D. Butcher as their farmhands
stood a respectful distance behind them on
a farm in western Nebraska, about 1903.
The Samson was manufactured by the
Stover Manufacturing Company of Freeport,
Illinois. Courtesy of the Nebraska State
Historical Society, Lincoln, Nebraska.

*Thousands of steel windmills on steel towers like this one dotted the level fields of the Midwest, one or more to each farm, at the beginning of the twentieth century. From the author's collection.*

*A springtime wedding party gathered on this Midwestern farm sometime during the early twentieth century. In the background is a steel vaneless windmill with an unusual, possibly homemade, counterbalance weight in the shape of a water pitcher. From the author's collection.*

Windmillers served the needs of farmers as well as ranchers. In this circa 1925 photograph, Minnesota windmiller Lucius Sweet posed with his daughter at the top of a fifty-foot steel tower supporting a twelve-foot-diameter Monitor Vaneless Style M windmill made by the Baker Manufacturing Company of Evansville, Wisconsin. He wrote by hand on the back of the picture, "On the platform is my little girl Sylvia and myself. She is 11 years old and has been helping me build towers for three summers during vacation. I have been giving her 25¢ per day but she thinks she should have 50 cts. this summer. She delights in climbing." From the author's collection.

Near Beatrice, Nebraska, where the Dempster Mill Manufacturing Company made this Dempster Steel windmill, two boys in the loft of a barn look out over a snowdrift that completely covered the base of the windmill tower in 1904. From the author's collection.

A Perkins Steel windmill made by the Perkins Wind Mill Company of Mishawaka, Indiana, supplied water to this snow-covered residence somewhere on the prairies about 1910. From the author's collection.

An Oklahoma Boomer windmill, patented by Rufus W. Smith of Oklahoma City in 1899, stood above this barnyard on the southern Great Plains at the turn of the twentieth century. From the author's collection.

J. G. Martin in his backyard in subtropical Fallbrook, California, around 1910. The lush vegetation was watered by a Woodmanse Steel windmill that had been produced over 1,000 miles away by the Woodmanse Manufacturing Company in Freeport, Illinois. From the author's collection.

*A windmill pumped water into this earthen stock-watering tank, where a Church of Christ minister administered baptism to Robert Hall at Petersburg, Texas, in the early twentieth century. Courtesy of the Southwest Collection, Texas Tech University, Lubbock, Texas.*

The Tuttle Grammar School, near Merced, California, received its water supply from a Pumping Aermotor that was mounted onto the side of a very distinctive wood-frame tank tower. Circa 1930. From the author's collection.

An IXL Steel windmill made by the Phelps and Bigelow Wind Mill Company of Kalamazoo, Michigan, elevated water into an insulated overhead tank behind this imposing early-twentieth-century rural home in the Pacific Northwest, described in handwriting on the back of the photograph as "out in Washington." From the author's collection.

This country home near Vaughn, Washington, received its water supply from a Pumping Aermotor. The handwritten inscription on the photograph reads, "This is our home fronting Puget Sound, taken in the winter. . . . Loda in the window of her room[,] Uncle L. D. outside." From the author's collection.

A young girl in a pony cart sits in front of a Victorian home that had its own ten-foot-diameter regular-pattern Eclipse windmill and large overhead storage tank somewhere on the central Great Plains at the turn of the twentieth century. From the author's collection.

Architect John Winford Byars designed this enclosed tank tower supporting a Steel Eclipse Type WG windmill made by Fairbanks, Morse and Company of Chicago for the rural home of Harold W. Tuttle near Las Turas Lake, California, during the 1920s. From the author's collection.

Specialized farmers also needed fresh water. The owner of Carr's Nurseries at Yellow Springs, Ohio, erected this vertical-axis horizontal windmill sometime around the turn of the twentieth century. The unusual-looking wheel inside the rectangular tower produced the motive force. Courtesy of David Huber.

These six young men could not resist the temptation to climb the tower for a Samson windmill on a farm probably in Hall County, Texas, about 1920. Little brother was left on the ground with the pet dogs, unable to climb where the big boys had gone. From the author's collection.

During the 1890s, windmill irrigation became an agricultural fad on the Great Plains and in the Central Valley of California. Here an open back-geared Pumping Aermotor elevated groundwater into an earthen reservoir on the E. L. Hall farm. Hall then released the impounded water to irrigate crops on his farm near Garden City, Kansas. Courtesy of the Kansas State Historical Society, Topeka, Kansas.

The owners of even the most expensive homes in rural areas far from an electrical power supply often used the wind to pump their water, as did the family residing in this beautiful Pueblo Deco–style dwelling near Glencoe, New Mexico, about 1927. They chose the Stover Oil-Rite made by the Stover Manufacturing and Engine Company of Freeport, Illinois. From the author's collection.

# CHAPTER 7

# Homemade Windmills

Not all windmills came from factories. A handful of those who used wind power constructed their own machines to use the free power of the wind. These devices were not necessarily as efficient as those from the factories, and they often lacked governors to regulate their speed to protect them during windstorms. They did compete successfully with factory-made windmills, however, on the basis of economy. Some of them were incredibly cheap.

The simplest of the homemade mills consisted of rectangular wooden boxes that had four, six, or eight paddle-shaped blades that were exposed to the wind where they stuck up above the edges of the boxes. They bore the same relation to air that an overshot waterwheel does to water. Known variously as go-devil, ground-tumbler, and jumbo mills, they were used throughout the plains and prairies.

Battle-axe mills were more sophisticated. These had towers similar to conventional windmills but used wheels comprised of large paddle-shaped blades. Both battle-axe and jumbo mills were built so the blades pointed toward prevailing winds and they had no governors. When the wind blew too strongly and the wheels turned too fast, these mills self-destructed from centrifugal force.

A number of farmers duplicated the principles they saw in factory-made windmills. They used wood and scraps of metal to replicate designs they observed in mass-produced machines.

A handful of the homemade mills had vertical axes and horizontal wheels. Some of these machines had circular metal tracks to support the blades, which usually were hinged on the outer ends. A few horizontal mills had shutters or hoods that allowed the wind to strike only some of the blades at one time.

*Two men inspect the homemade windmill constructed by J. L. Brown at the Midway Nurseries at Kearney, Nebraska, in 1898. The mill measured only three feet wide but was nine feet long and six feet high. It was made from the sides and ends of discarded wooden packing crates and used a piece of gas pipe as its axle. Brown spent only $1.50 to build it, but it pumped enough water to irrigate a truck garden of vegetables and strawberries through an entire summer of drought. Photo by Erwin Hinckley Barbour. Courtesy of the U.S. Geological Survey Photo Archives, Denver, Colorado.*

During the 1890s, J. S. Peckham built these two giant battle-axe homemade windmills to irrigate a fifteen-acre vineyard and orchard. He constructed the towers, which supported wind wheels sixteen feet in diameter, from 4 × 4 and 2 × 4 lumber. When Professor Erwin Hinckley Barbour from the University of Nebraska visited in 1898, he wrote, "They were admirably constructed and ran as smoothly and noiselessly as a steel mill." Photo by Erwin Hinckley Barbour. Courtesy of the U.S. Geological Survey Photo Archives, Denver, Colorado.

During the 1930s, another style of homemade windmill appeared. It also consisted of second-hand components, cast-off 55-gallon steel drums and the differential gears from wrecked motor vehicles. Builders turned the differentials upside down at the tops of towers, where they became vertical axes that supported steel drums that were cut in half and mounted on arms.

The appeal of using free power from the wind was contagious in some neighborhoods, and some farmers and ranchers matched wits to create the best homemade wind machines. These grassroots inventions demonstrate the ingenuity of rural people in using the resources readily at hand.

Some people who built homemade
windmills simply made crude copies of
factory-made windmills. This is what Henry
Boersen did on his farm near Grand Island,
Nebraska, in the 1890s. He adapted an old
wooden pump to form the body of his mill,
taking most of the other parts from a
disused corn sheller. Boersen spent only
$2.00 to build this mill, which watered
sixty head of cattle. Photo by Erwin Hinckley
Barbour. Courtesy of the U.S. Geological
Survey Photo Archives, Denver, Colorado.

In the 1890s, Mr. S. S. Videtto near Lincoln,
Nebraska, made a different style of
homemade windmill. His had a vertical axis
and a series of pivoted blades that were
free on their outer edges. The wheel spun
like a merry-go-round, providing motive
power for a pump mechanism. Photo by
Erwin Hinckley Barbour. Courtesy of the U.S.
Geological Survey Photo Archives, Denver,
Colorado.

*Two men stand beside a large wheel for a homemade windmill, possibly in California. Early twentieth century. From the author's collection.*

In eastern Canada at the turn of the twentieth century, a number of people built four-bladed windmills on the general principle of this mill, which was used to cut firewood. As the wheel turned, the machine moved a large bow saw back and forth to cut timber into usable sizes. From the author's collection.

Using the staves from a wooden barrel to form the blades, Howard Flickinger, a farmer near Skiff, Alberta, constructed this homemade windmill about 1915 to spin a keg-style butter churn. He mounted the machine on a sledge so he could move it around or turn it to face the wind. Courtesy of the Canadian National Historic Windmill Centre.

Truman Chandler, a young man who migrated to Merced, California, from Oklahoma during the days of the Dust Bowl, constructed this homemade windmill that used two steel drums cut in half and the differential from a wrecked automobile to pump water at his modest home. Agricultural engineer James P. Fairbank made this photograph of Chandler on the porch of his home beside the windmill on July 2, 1941. Courtesy of the Agricultural Engineering Archives (AR 012), Special Collections, University of California Library, Davis, California.

# WIND-DAMAGED WINDMILLS

Windmills are like watermills. They use free power provided by nature, but that power at times proves too strong for the machines to handle. Watermills wash away in floods, and windmills tear apart in the wind.

Damage comes from several different types of wind. Sometimes the wind is simply so strong, as in a hurricane, that it blows down or damages anything that rises up from the surface of the land. But windmills were designed by engineers to survive most winds under hurricane force. If they are erected properly with sufficient anchoring and appropriately maintained, most windmills survive straightforward high-velocity wind.

Wind turbulence is the great destroyer of windmills. If an owner builds a windmill in a location prone to turbulence, such as the entrance to a canyon or the base of a mountain range, natural turbulence may lead to the destruction of the mill. Even more serious is the problem of windmills erected near obstructions like houses, barns, and trees. Such mills and their towers are subject to the turbulence that occurs as the wind blows around such obstacles, and they frequently fail.

A final source of turbulence for windmills comes from tornadoes. The extreme low barometric pressure of these storms causes winds to swirl at great speed. No windmill or tower is built to withstand such force, and frequently one sees windmills in the field so twisted from tornado winds that the remains are almost unrecognizable.

Severe winds downed buildings and a windmill tower on the Sam Haines farm somewhere on the plains in the early twentieth century, dashing the spirits of the somber couple who survived the storm. From the author's collection.

Burdick & Burdick Company employees in the desert Southwest pose with pieces of the wheel from a Challenge 27 windmill that was torn to pieces during a windstorm about 1935. Courtesy of B. H. Burdick, Sr.

# CHAPTER 8

# Windmills and Railroads

One of the most important requirements in the operation of steam railways was a clean water supply. Steam locomotives needed water to run, and it had to be clean to avoid building up scale from mineral deposits inside the boiler tubes. Locomotives stopped repeatedly to take on replacement water as they traveled. This meant that railway companies spent substantial money to provide clean water at appropriate intervals along their tracks.

In the East, railroads generally were able to find surface water easily. As they laid rails westward onto the Great Plains and beyond, though, the land grew more and more arid. Surface water was often unavailable, and the cost of buying coal and then hauling it to remote locations to fuel steam pumps was prohibitive. Windmills offered a viable option for elevating the water that railroads needed.

By the 1870s, railway companies had begun to use windmills to pump groundwater for their steam engines. In response, manufacturer after manufacturer began to produce large-diameter heavy-duty mills termed "railroad pattern." The most popular of the early mills for locomotive water

This large Halladay Standard windmill made by the U.S. Wind Engine and Pump Company of Batavia, Illinois, pumped boiler water for steam locomotives on the Union Pacific Railroad at Laramie, Wyoming, in 1870. Photo by J. W. Jackson. Courtesy of the Union Pacific Historical Collection, Council Bluffs, Iowa.

*This unidentified large-diameter railroad-pattern wooden-wheel windmill over an unusual tank tower pumped boiler water alongside railway tracks on a stream in the eastern United States at the end of the nineteenth century. From the author's collection.*

*A large-diameter Halladay Standard windmill on an unusually short tower pumped water for Union Pacific steam locomotives at Sherman, Wyoming. Photographed by J. W. Jackson in 1870. Courtesy of the Union Pacific Historical Collection, Council Bluffs, Iowa.*

supply was the Railroad Eclipse made by the Eclipse Wind Mill Company, the Eclipse Wind Engine Company, and finally by Fairbanks, Morse and Company. These were sturdier and larger than most windmills; they came in sizes up to thirty feet in diameter.

Many of the manufacturers also produced or sold large wood-stave water tanks that came with special spouts to fill locomotive tenders quickly. The mills pumped most of the time so the tanks lining the railway could be ready at any time to water locomotives. The large-diameter mills and elevated tanks were common sights along railways in many parts of the country during the late nineteenth and early twentieth centuries. Some municipalities also used railroad-pattern windmills for domestic water supply, and some ranches used the big mills for deep or large-volume pumping for livestock.

Courtesy Mr. Wm. Reinhardt, Industrial Agt, Los Angeles

A steam locomotive pulled forward from the station at Dows, Iowa, in the early twentieth century with a Railroad Eclipse windmill made by Fairbanks, Morse and Company of Chicago in the background. From the author's collection.

A Railroad Eclipse windmill stood beside grain elevators and a railway water tank alongside the tracks in Syracuse, Nebraska, in the early twentieth century. The tower for this mill had an exceptionally large circular service platform. From the author's collection.

A view, possibly from the top of a grain elevator, showing a Railroad Eclipse windmill and large overhead tank for watering steam locomotives on the prairie at South Shore, South Dakota, in the early twentieth century. From the author's collection.

A Railroad Eclipse windmill elevated water into a wood-stave water tank for railway locomotive use at Pilger, Nebraska, about 1906. The large pivoted spout conducted water from the tank into locomotive tenders quickly and efficiently as the windmill gradually refilled the tank for the next engine. From the author's collection.

This Railroad Eclipse windmill pumped water for locomotives at Niagara, North Dakota, in the early twentieth century. From the author's collection.

A Railroad Eclipse windmill and wood-stave water tank beside the railway tracks in Creighton, Nebraska, about 1911. The rail car beside the tank was designed to transport new automobiles. From the author's collection.

A huge Railroad Eclipse windmill pumped water beside the tracks of an unidentified town at the turn of the twentieth century. From the author's collection.

*General view of grain elevators, Railroad Eclipse windmill, elevated railway water tank, and grain-harvesting machinery at Plaza, North Dakota, in the early twentieth century. From the author's collection.*

*View down the tracks toward Ripon, Wisconsin, and its windmill water system for steam locomotives about 1909. From the author's collection.*

*The effects of frost on the railroad water system at Judd, Wisconsin, in the early twentieth century. A man at its base sits on a hand-operated railway velocipede on the tracks. From the author's collection.*

# WINDMILLS OVER PUBLIC WELLS

Throughout the nation, local governments funded the sinking of wells and often the construction of windmills for public use. The horses and mules people relied on for transportation before the automobile era required regular access to water.

Frequently a municipality or county placed a public well in the center of a main thoroughfare so they would be easy to get to without crowding. Next to the wells they constructed concrete, wooden, or steel troughs with sides the appropriate heights for animals to drink comfortably. The water systems generally were equipped with float valves and windmill regulators so the mechanisms turned off when the tanks filled but then turned back on when animals or evaporation had lowered the level. This helped keep the area around such wells from becoming muddy bogs.

Many historic photographs illustrate public wells equipped with windmills. The images began appearing in the late nineteenth century and were common until the beginning of the automobile era in the second decade of the twentieth century, after which time they mostly disappeared.

*Workers were completing the water troughs at the base of this Fairbury No. 1 Vaneless windmill made by the Fairbury Iron Works and Windmill Company of Fairbury, Nebraska, when the photographer took this picture on December 24, 1908. The windmill was located in the center of Main Street at Ruskin, Nebraska. From the author's collection.*

A Fairbanks-Morse Steel mill made by Fairbanks, Morse and Company and sold through its branch house in Kansas City graced the principal thoroughfare of this Kansas town, where horses could drink from a concrete trough. Circa 1910. From the author's collection.

Farnam, Nebraska, boasted an impressive Eclipse windmill in its main street that gave locals and travelers convenient access to water in the early twentieth century. Courtesy of the Nebraska State Historical Society, Lincoln, Nebraska.

# CHAPTER 9

# Windmills in Urban Settings

**A**s soon as windmills began appearing on farms and ranches, people started putting them to work in urban settings. Initially individuals erected windmills to pump groundwater from drilled or hand-dug wells at their homes and places of business. Then municipalities started placing windmills over public wells to provide clean water for residents and visitors. Eventually some communities constructed entire waterworks systems that used windmills to pump water into central reservoirs and then through pipes to consumers in the towns.

In time, communities constructed more-sophisticated waterworks systems and power pumps supplanted windmills when the demand for water exceeded the supply traditional windmills could provide. Yet many individuals kept windmills and private wells instead of using city water or as backup sources in case the central systems broke down. Windmills are still found in towns in many parts of the United States.

Resorts transplanted urban life to scenic rural settings. From the mountaintops to the seashores, developers of these facilities found that windmills were a practical means of pumping water into elevated reservoirs. The pressure of gravity forced water from the tanks through pipes so guests and employees could use it for drinking and washing. In some instances, resort operators also used power windmills, which produced rotary motion in steel shafts, to grind grain for horses and other animals, and in later years they used wind generators to produce limited amounts of electricity.

Both urban and rural dwellers enjoyed watching the spinning wheels of miniature windmills. Builders and users of these models ranged from children to adults. Bona fide windmill manufacturers produced precise miniatures of their working windmills for traveling salesmen to use to demonstrate to prospective customers how the machines worked. These firms also made somewhat larger models as gifts for especially valuable dealers, who installed them on sales counters or salesroom floors. The resort areas of Cape Cod became famous for their handmade miniature windmills, which travelers took back home to cities and towns throughout the United States. All across America, businesspeople erected miniature windmills of all sizes to catch the eye of potential customers.

*These two ladies must have enjoyed climbing to the top of the tower for this fourteen-foot-diameter Original Star windmill made by the Flint and Walling Manufacturing Company of Kendallville, Indiana. From their vantage point they could look over most of the city of San Diego, California, about 1880. Courtesy of the San Diego Historical Society, San Diego, California.*

These three big Woodmanse Mogul windmills made by the Woodmanse Manufacturing Company of Freeport, Illinois, pumped the water supply for Grant, Nebraska, during the early twentieth century. Behind them is the riveted-steel standpipe into which they elevated water. From the author's collection.

Several varieties of windmills dotted the landscape around this courthouse square in West Texas during the early twentieth century. Courtesy of the Panhandle-Plains Historical Museum, Canyon, Texas.

This Virginia windmill made by the Wise Wagon Works of Buena Vista, Virginia, stood atop an ornamental enclosed tank tower that provided water for this residence somewhere in the eastern United States during the 1890s. From the author's collection.

*The wheel on this windmill atop a house in San Francisco was comprised of paddle-shaped blades. In the background is the Golden Gate entry to San Francisco Bay. Alcatraz Island is on the right. This photo was taken in 1875. Courtesy of the California State Library, Sacramento, California.*

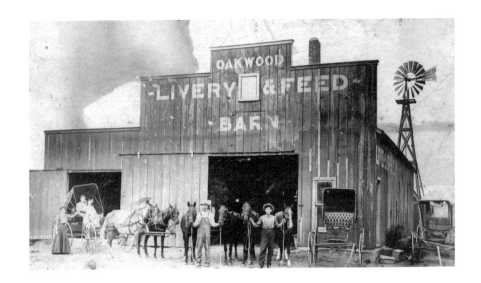

*A Monitor Steel windmill made by the Baker Manufacturing Company of Evansville, Wisconsin, pumped water for the Oakwood Livery and Feed Barn, possibly in California, in the early twentieth century. From the author's collection.*

*A Railroad Eclipse windmill made by Fairbanks, Morse and Company of Chicago was reflected in the glass store windows as a photographer made this picture of the "all girl" band in Durand, Illinois, in the early twentieth century. From the author's collection.*

A large-diameter Railroad Eclipse windmill on a steel tower dominated the townscape as it pumped the municipal water supply for Gimli, Manitoba, in the late 1930s. Courtesy of the Archives of Manitoba, Winnipeg, Manitoba.

This battery of Challenge 27 windmills made by the Challenge Company of Batavia, Illinois, and Samson Oil-Rite windmills made by the Stover Manufacturing and Engine Company of Freeport, Illinois, kept the Cocklin Fish Farm reservoirs filled in Griswold, Iowa, during the 1920s. From the author's collection.

Wind-power experimenter Oliver Parker Fritchle combined this Woodmanse Solid Wheel windmill made by the Woodmanse Manufacturing Company of Freeport, Illinois, with a dynamo to generate electricity for home use in Colorado. He stood alongside his battery-powered electric automobile at the Warminster, Colorado, residence of Jay Hammontree about 1920. Courtesy of the Colorado Historical Society, Denver, Colorado.

# WINDMILLS AT RESORTS

When they developed resorts, entrepreneurs attempted to re-create the convenience of urban living in appealing rural settings. Their goal was to provide guests with the beauty of nature and the comforts of the city. Windmills played important roles in these business enterprises.

Frequently builders and managers of resorts chose windmills to pump water for their facilities. They generally placed mills over wells to pump water from under the ground to elevated reservoirs made of wood or metal. Then they built piping systems that carried the water to their guests and employees, who could use faucets to control the indoor water supply just the same as in city waterworks systems.

Resort windmills were built in settings that ranged from seashores to mountains, and windmills frequently appeared in historic photographs made in such scenic environments.

*An Iron Turbine, the first commercially successful all-metal windmill in the United States, made by Mast, Foos and Company of Springfield, Ohio, pumped water for the Summit House atop Mount Wachusett, Massachusetts, in the 1880s. From the author's collection.*

A Freeman Steel power windmill made by S. Freeman and Sons Manufacturing Company of Racine, Wisconsin, ground grain for animal feed at a log cabin resort somewhere in the American West during the early twentieth century. From the author's collection.

A ten-foot-diameter regular-pattern Eclipse windmill pumped water into a square masonry reservoir for travelers at this roadside auto campground during the 1930s. From the author's collection.

Mounted on the roof of the main building, a Wincharger wind generator made by the Wincharger Corporation of Sioux City, Iowa, produced six-volt electric power for a few light bulbs and probably a radio receiver at this campground near Taos, New Mexico, during the 1930s. From the author's collection.

# MINIATURE WINDMILLS

Adults and children alike seem to share a love for miniature windmills. Everyone seems to enjoy watching tiny blades spinning around even if they actually perform no work.

Because of this wide appeal, miniature windmills frequently show up in historic photographs. The models may be handmade toys for children or actual scale models produced by or for bona fide windmill makers to demonstrate the features of their mass-produced goods.

Entrepreneurs took advantage of the appeal of miniature windmills; they placed them in shop windows and erected them in front of their places of business to attract customers. Others created business enterprises by making and selling miniature windmills. Cape Cod became famous in the early twentieth century for its "windmill shops" and "mill works" that specialized in toy windmills and whirligigs.

*This lucky boy stands beside a homemade miniature windmill complete with a simulated pump at Berrien Center, Michigan, in the 1880s. From the author's collection.*

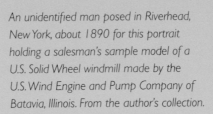

*An unidentified man posed in Riverhead, New York, about 1890 for this portrait holding a salesman's sample model of a U.S. Solid Wheel windmill made by the U.S. Wind Engine and Pump Company of Batavia, Illinois. From the author's collection.*

132

The appeal of miniature windmills led several individuals, among them Atherton Crowell of Dennis Port, to produce them for the tourist trade on Cape Cod, Massachusetts, during the early twentieth century. From the author's collection.

These two gentlemen of the Oklahoma Territory around the turn of the twentieth century posed for a studio portrait in front of a miniature windmill in Oklahoma City. Courtesy of the Panhandle-Plains Historical Museum, Canyon, Texas.

Miniature windmills served as eye-catchers for business establishments, like this stylized example in front of the Blue Bonnet Auto Court in Santa Rosa, California, during the 1920s. From the author's collection.

# Windmills Overseas

**A**s early as the 1860s, American companies began exporting factory-made windmills to overseas markets. At the same time, they also exported the concepts behind the machines. Concurrently scientists in other countries investigated methods of using the wind more efficiently.

American-made windmills found markets around the world. By the boxcar they went to ports for ocean shipment to other lands. To this day, U.S.-made windmills survive by the thousands in such regions as southern Africa and Australia.

No sooner did dealers start selling imported windmills than local businesspeople saw that money could be made producing similar machines locally. People like James Alston and George Washington Griffiths in Australia, Auguste Bollée in France, Carl Herzog in Germany, and John Wallis Titt in England experimented with wind power, producing their own designs for wind machines that they too exported outside their home countries. In this way, factory-made turbine-wheel windmills came to be used, to at least some extent, in most countries around the world where the wind could be exploited to lift water and do other human work. Today the largest windmill factory in the world is located near Buenos Aires, Argentina.

*Steel windmills, including a Samson made by the Stover Manufacturing Company of Freeport, Illinois, visually dominated this exhibition of agricultural machinery at Sousse, Tunisia, in 1910. From the author's collection.*

CONCOURS DE SOUSSE (1910)
PHOTO MORELLI

*Throughout North Africa, starting in the late nineteenth century, factory-made windmills created artificial oases where people and animals could reliably find water. Here a windmill pumped water early in the twentieth century to supply a lead and zinc mine at the foot of the mountain called Djebel Ressas in the El Kef area of Tunisia. From the author's collection.*

*This large Halladay Standard windmill made by the U.S. Wind Engine and Pump Company of Batavia, Illinois, stood atop a building facing the Indian Ocean at Port Dauphin on the island of Madagascar in the early twentieth century. From the author's collection.*

These horseback riders knew they could find fresh water in the Australian outback at this eight-foot-diameter Alston Double Crank steel windmill made by James Alston and Sons, Ltd., of Melbourne, Australia, in the early twentieth century. From the author's collection.

This big oil-bath Alston steel windmill
pumped water for both sheep and cattle
on Kangaroo Island, just off the south coast
of Australia, around 1935. From the author's
collection.

Founded by George Washington Griffiths in
1871, the Toowoomba Foundry Company,
Ltd., and its successors have manufactured
windmills in Australia for over 125 years. The
seventeen-foot-diameter Southern Cross R
Pattern windmill pictured here was erected
in an open savannah setting around 1980.
Courtesy of Southern Cross Pumps and
Irrigation, Pty. Ltd., Withcott, Queensland,
Australia.

The Williams Manufacturing Company of Kalamazoo, Michigan, exported its Manvel solid-wheel wooden windmills to markets around the world. This Manvel was mounted on a tubular wooden mast above a barn in England sometime around the turn of the twentieth century. Courtesy of J. Kenneth Major.

People applied the concept of the solid-wheel wooden windmill to their local conditions, as occurred at Palma de Mallorca, Spain, in the early twentieth century. From the author's collection.

Following the lead of scientist Poul la Cour, Danes adapted wind power technology to produce their own distinctive wind machines. Windmills like this one, which uses variable-pitch blades and a fan tail to direct the wheel into the wind, were used widely in the Danish countryside during the first decades of the twentieth century. From the author's collection.

Hercules windmills, which were manufactured in Dresden, Germany, during the first four decades of the twentieth century, were used in many European countries. From the author's collection.

Soldiers climbed onto the steel tower of this wooden windmill with a sectional wheel somewhere in France around the time of World War I. From the author's collection.

PARQUE, VISTA DAS FONTES    CAMBUQUIRA

*Throughout South America, turbine-wheel windmills were used to pump water for both people and animals, as in this park in Cambuquira, Brazil, in the early twentieth century. From the author's collection.*

The shallow depth of the fresh-water aquifers in the Caribbean islands made it possible for people there to use factory-made windmills from the United States to provide their supply. On the island of Curaçao, a Dandy made by the Challenge Company of Batavia, Illinois, and a Star Model 24 made by the Flint and Walling Manufacturing Company of Kendallville, Indiana, pumped water for local people around 1930. From the author's collection.

*This Star Model 12 made by the Flint and Walling Manufacturing Company provided water for the oceanside Casa Piño resort in Cuba about 1930. From the author's collection.*

POR LAS AZOTEAS DE CAMPECHE CAMP.MEX.

At Campeche, Mexico, a veritable forest of factory-made American windmills pumped water for many of the residents of the city during the first half of the twentieth century. From the author's collection.

# Index